建筑与都市系列丛书 | 世界建筑
Architecture and Urbanism Series | World Architecture

文筑国际 编译
Edited by CA-GROUP

Architecture in Chile:
In Search of a New Identity

智利：新身份探求

中国建筑工业出版社

图书在版编目（CIP）数据

智利：新身份探求 = Architecture in Chile:In Search of a New Identity：英汉对照 / 文筑国际 CA-GROUP 编译 . —北京：中国建筑工业出版社，2021.1
（建筑与都市系列丛书 . 世界建筑）
ISBN 978-7-112-25736-2

Ⅰ . ①智 ... Ⅱ . ①文 ... Ⅲ . ①建筑艺术 - 介绍 - 智利 - 英、汉 Ⅳ . ① TU-867.84

中国版本图书馆 CIP 数据核字 (2020) 第 254197 号

责任编辑：毕凤鸣　刘文昕
版式设计：文筑国际
责任校对：芦欣甜

建筑与都市系列丛书 | 世界建筑
Architecture and Urbanism Series | World Architecture
智利：　新身份探求
Architecture in Chile: In Search of a New Identity
文筑国际　编译
Edited by CA-GROUP

*

中国建筑工业出版社出版、发行（北京海淀三里河路 9 号）
各地新华书店、建筑书店经销
北京雅昌艺术印刷有限公司 制版、印刷

*

开本：787 毫米 ×1092 毫米　1/16　印张：15½　字数：449 千字
2024 年 7 月第一版　　2024 年 7 月第一次印刷
定价：**148.00** 元
ISBN 978-7-112-25736-2
（36505）

版权所有　翻印必究
如有内容及印装质量问题，请联系本社读者服务中心退换
电话：　(010) 58337283　QQ：2885381756
（地址：北京海淀三里河路 9 号中国建筑工业出版社 604 室　邮政编码 100037）

a+u

建筑与都市系列丛书学术委员会
Academic Board Members of Architecture and Urbanism Series

委员会顾问 Advisors
郑时龄 ZHENG Shiling　崔 愷 CUI Kai　孙继伟 SUN Jiwei

委员会主任 Director of the Academic Board
李翔宁 LI Xiangning

委员会成员 Academic Board
曹嘉明 CAO Jiaming　张永和 CHANG Yungho　方 海 FANG Hai
韩林飞 HAN Linfei　刘克成 LIU Kecheng　马岩松 MA Yansong
裴 钊 PEI Zhao　阮 昕 RUAN Xing　王 飞 WANG Fei
王 澍 WANG Shu　赵 扬 ZHAO Yang　朱 锫 ZHU Pei

*委员会成员按汉语拼音排序（左起）
Academic board members are ranked in pinyin order from left.

建筑与都市系列丛书
Architecture and Urbanism Series

总策划 Production
国际建筑联盟 IAM　文筑国际 CA-GROUP

出品人 Publisher
马卫东 MA Weidong

总策划人/总监制 Executive Producer
马卫东 MA Weidong

内容担当 Editor in Charge
吴瑞香 WU Ruixiang

助理 Assistants
卢亭羽 LU Tingyu　杨 文 YANG Wen　杨紫薇 YANG Ziwei

特约审校 Proofreaders
寇宗捷 KOU Zongjie

翻译 Translators
英译中 Chinese Translation from English:
樊梦莹 FAN Mengying (pp.12-p23)　林天成 LIN Tiancheng (pp.28-p229)
日译中 Chinese Translation from Japanese:
吴瑞香 WU Ruixiang (p.11)

书籍设计 Book Design
文筑国际 CA-GROUP

中日邦交正常化50周年纪念项目
The 50th Anniversary of the Normalization of
China-Japan Diplomatic Relations

本系列丛书部分内容选自A+U第594号（2020年03月号）特辑
原版书名：
チリの建築——新たなアイデンティティの探求
Architecture in Chile: In Search of a New Identity
著作权归属A+U Publishing Co., Ltd. 2020

A+U Publishing Co., Ltd.
发行人/主编：吉田信之
副主编：横山圭
内容担当：服部真吏　Sylvia Chen
海外协助：侯 蕾

Part of this series is selected from the original a+u No. 594 (20:03),
the original title is:
チリの建築——新たなアイデンティティの探求
Architecture in Chile: In Search of a New Identity
The copyright of this part is owned by A+U Publishing Co., Ltd. 2020

A+U Publishing Co., Ltd.
Publisher / Chief Editor: Nobuyuki Yoshida
Senior Editor: Kei Yokoyama
Editorial Staff: Mari Hattori Sylvia Chen
Oversea Assistant: HOU Lei

封面图：阿塔卡马沙漠博物馆室外夜景，建筑设计：柯兹，波里杜拉，博兰德，索托
封底图：玻璃房室内，建筑设计：马克斯·努涅斯

本书第232页至248页内容由安藤忠雄建筑研究所与和美术馆提供，在此表示特别感谢。

本系列丛书著作权归属文筑国际，未经允许不得转载。本书授权中国建筑工业出版社出版、发行

Front cover: Exterior evening view of Museum of the Atacama Desert, designed by Coz, Polidura, Volante, Soto. Photo by Sergio Pirrone.
Back cover: Interior view of the Glass House, designed by Max Núñez. Photo by Roland Halbe.

The contents from pages 232 to 248 of this book were provided by Tadao Ando Architect & Associates and He Art Museum. We would like to express our special thanks here.

The copyright of this series is owned by CA-GROUP. No reproduction without permission. This book is authorized to be published and distributed by China Architecture & Building Press.

Preface:
Wandering between Mountain and Ocean

PEI Zhao

Chile's unique geography has been shaped over millions of years by the collision of American and Pacific tectonic plates, a land where the vertical Andes and the horizontal Pacific meet, with frequent earthquakes, extreme climates abound, and the people who live between the permanence of nature and the uncertainty of real world. "How can people who live between such ocean and such mountain range without poetry?" Hence poetry became a way for Chileans to enter their lives and environments, also bring some poetic dimension for Chilean architecture.

As early as 1855, almost the first architectural textbook in Latin America is published in Chile; at the end of the 19th century, while other Latin American countries were busy on fighting over architectural styles, Chile had established a complete architectural education system imitating Beaux-Arts of Paris; in the 1930s, when the Brazilian and Mexican modernists struggled, the ideas modernism architecture was quietly introduced into Chilean university. Chile's systematic professional education in architecture has led to a very different style of thinking and practice, far from that of other Latin American countries. Even during Golden Age of Latin American modernism architecture (1930-1950), Chilean architecture seems to be hesitated without joining in the carnival of such nation grand narrative.

In the 1960s, Chile saw the emergence of modern architectures with international repute; more importantly, the emergence of Open City in Valparaíso provided alternative path for Latin American modernism architecture by seeking cultural identity not based on national ideology, but from the collective and the individual lives, from poetry and art. Although these explorations were interrupted by the military junta's rise to power and then lost in postmodernism, such accumulation was eventually unleashed in the 1992 Seville Expo, where the Chilean and unbuilt Brazilian pavilions became symbols of the decline of postmodernist architecture in Latin America.

Since then, a wave of Chilean architects and buildings have appeared on the international scene. They are not trapped by the strong architectural traditions of Latin America, stepped out of the shadow of the old masters, and abandoned the deliberate pursuit for "Latin America" and "Chilean-ness", and instantaneously responded to the eternal nature and the unpredictable reality of the society with personal and diversified design ways, resulting in contemporary Chilean architecture that is both poetic and highly realistic.

To explain such unique mental of contemporary Chilean architecture, we may borrow the words of an ever-wandering poet, a figure in Roberto Bolaño's *The Savage Detectives*, "We weren't writing for publication but to understand ourselves better or just to see how far we could go."

序言：
徘徊于山海之间

裴钊

千万年来美洲板块与太平洋板块的碰撞塑造了智利独特的地理。在垂直的安第斯山脉和水平的太平洋交汇处，地震频繁，充满各种极端气候，人们生活在山海的永恒与现实环境的不确定之间。"夹在这样的海洋和这样的山脉之间生活的人们怎么能没有诗歌？"因此，诗歌成为智利人进入生活和环境的一种方式，也渲染了智利的建筑。

早在 1855 年，智利就出版了几乎是拉美的第一本建筑教科书；19 世纪末，当其他拉美国家忙于建筑风格之争时，智利已经仿照巴黎美院建立了完整的建筑教育体系；1930 年代，当巴西和墨西哥的现代主义者奋力抗争时，现代主义建筑理念悄然地被引入智利的大学教学之中。系统的建筑职业教育使得智利的建筑思考和实践呈现出与拉美其他国家迥异的风格，即使在拉美现代主义建筑黄金时代（1930-1950 年），智利建筑也似乎在犹豫，没有加入国族宏大叙事的狂欢。

1960 年代，智利出现了一批具有国际声誉的现代建筑杰作；更为重要的是瓦尔帕莱索（Valparaíso）开放城市的出现，为拉美现代主义建筑开辟了另外一条道路，不是基于国家意识形态寻求文化认同，而是从集体和个体出发，从诗歌和艺术出发，在建筑中寻找文化认同。尽管这种探索因为军政府的上台而中断，也因后现代主义而迷失，但其积蓄的力量最终在 1992 年塞维利亚博览会中得到了释放，智利馆与未建成的巴西馆成为后现代主义建筑在拉美式微的象征。

此后，智利优秀的建筑师和建筑如浪潮一般出现在国际视野中。他们没有受困于拉美强大的建筑传统，走出了老一辈天才大师的阴影，也摒弃了对"拉美性"和"智利性"的刻意追求，以极其个人和多元的设计，来即时响应永恒的自然和变幻无常的现实社会，产生出既具有诗意、又高度现实主义的当代智利建筑。

智利当代建筑所展出的这种独特心智，或许可以借用智利文学家罗贝托·波拉尼奥（Roberto Bolaño）的小说《荒野侦探》中一位流浪诗人的话来解释：

"我们创作并不是为了发表，而是为了更好地理解自我，或者只是想看看我们能走多远。"

PEI Zhao

Research scholars on Latin American architecture. Executive director of Latin American Research Center of China Architectural Culture Research Association. Guest professor of School of architecture, University of Miami.

裴钊

拉美建筑研究学者、中国建筑文化研究会拉美研究中心执行主任、迈阿密大学建筑学院客座教授

Architecture in Chile:
In Search of a New Identity

Preface:
Wandering between Mountain and Ocean   6
PEI Zhao

Introduction:
CHILE #2 – Architecture since 1990   12
Diego Grass

Sebastian Irarrazaval
Casa 2Y   28

Smiljan Radić
House for the Poem of the Right Angle   40
Prism House + Terrace Room   52

Pezo von Ellrichshausen
Loba House   64

Max Núñez
Ghat House   74

Guillermo Acuña
Isla Lebe   84

Ortuzar Gebauer Architects
Palafito del Mar Hotel   100

Cazú Zegers Architects
Hotel of the Wind – Tierra Patagonia Hotel   110

Coz, Polidura, Volante, Soto
Museum of the Atacama Desert   124

Undurraga Devés Architects
Retiro Chapel   134

Emilio Marín, Juan Carlos López
Desert Interpretation Center   146

Max Núñez
Glass House   154

Undurraga Devés Architects
Ruca Dwellings   166

OWAR Architects
Housing Complex in Quinta Normal   172

Izquierdo Lehmann Architects
Cruz del Sur Building   180

**Cristián Fernández Architects,
Lateral Architecture & Design**
Gabriela Mistral Cultural Center   190

Joaquín Velasco Rubio
Dinamarca399   200

Cecilia Puga, Paula Velasco, Alberto Moletto
Palacio Pereira   212

Architects Profile   224

Spotlight:
He Art Museum   232
Tadao Ando Architect & Associates

智利：
新身份探求

序言：
徘徊于山海之间　7
裴钊

引言：
智利 2.0——1990 年之后的建筑　12
迭戈·格拉斯

塞巴斯蒂安·伊拉拉萨瓦
2Y 之家　28

史密里安·拉迪奇
《直角之诗》之家　40
棱镜屋 + 露台屋　52

佩索·冯·艾奇绍森
罗巴住宅　64

马克斯·努涅斯
盖特住宅　74

吉列尔莫·阿库尼亚
莱贝岛　84

奥尔图扎尔·格鲍尔建筑师事务所
帕拉非托德马尔酒店　100

凯泽·塞赫尔斯建筑师事务所
风之酒店——蒂拉·巴塔哥尼亚酒店　110

柯兹，波里杜拉，博兰德，索托
阿塔卡马沙漠博物馆　124

翁都拉卡德维斯建筑师事务所
雷蒂罗礼拜堂　134

埃米利奥·马林，胡安·卡洛斯·洛佩斯
沙漠文化中心　146

马克斯·努涅斯
玻璃房　154

翁都拉卡德维斯建筑师事务所
卢卡住宅　166

OWAR 建筑师事务所
金塔诺马尔集合住宅　172

伊斯基尔多·莱曼建筑师事务所
克鲁兹·德尔·苏尔大楼　180

克里斯蒂安·费尔南德斯建筑师事务所，横向建筑师事务所
加夫列拉·米斯特拉尔文化中心　190

约阿奎因·贝拉斯科·鲁比奥
丹麦 399 号　200

塞西莉亚·普加，宝拉·贝拉斯科，阿尔贝托·莫莱托
佩雷拉宫　212

建筑师简介　224

特别收录：
和美术馆　232
安藤忠雄建筑研究所

Editor's Word

编者的话

Located at the end of the world, its idyllic landscape creates a perfect canvas for Chilean architects to express poetry in their architecture. To many, it is the utopian holiday homes that brought Chilean architecture into the international scene, and examples of these houses were previously featured in a+u 06:07 and in this book, the House for the Poem of the Right Angle (see pp. 40–51) and Loba House (see pp. 64–73). Following 2010, however, we begin to see a different group of architects looking into less individualistic visions. Guided by a moral compass, they engage with the public or take up non-profit projects – Ruca Dwellings (see pp. 166–167) and Gabriela Mistral Cultural Center (see pp. 190–199) – that focus on social and sustainable issues which came to a halt during the time of oppression. In an introductory essay, Diego Grass, an architect and tutor at Pontifical Catholic University of Chile, sets down to share with us his insights into Chilean architecture from the 1990's. He describes how having gone through years of persistent domestic unrest, the country seeks to forge a new cultural identity that would bring a divided Chile together. 18 projects are selected in this book to broaden our perspectives into architecture found in Chile, and the many ways these architects respond to its landscape and urban territory contexts. (a+u)

以"天涯之国"——智利素朴美丽的风土为画布，智利建筑师们创造了许多富有诗意的建筑，并获得了国际赞誉。无论是a+u在2006年曾编辑的日英版智利专辑《a+u06:07》，还是在本书中，a+u都收录了"《直角之诗》之家"（见本书40-51页）和"罗巴住宅"（见本书64-73页）等乌托邦式的假日住宅作品。而另一方面，在2010年之后，建筑师们与行政机关、非营利性团体机构一起致力于社会性、环境性的主题，这些都促使了一种建筑设计新潮的逐渐诞生。如收录在本书中的"卢卡住宅"（见本书166-171页）、加夫列拉·米斯特拉尔文化中心（见本书190-199页）等建筑作品。为本书撰写序文的是执教于智利天主教大学的建筑师迭戈·格拉斯，据迭戈所述，1990年后的智利建筑史，是智利从因长年动荡不安而致分裂的本国文化身份，向一种新的统一的身份逐渐摸索的轨迹。本书中，我们将通过与自然和城市相呼应的18个作品，向大家介绍智利的当下。

(a+u)

Introduction:
CHILE #2 – Architecture since 1990
Diego Grass

引言：
智利2.0——1990年之后的建筑
迭戈·格拉斯

1990–2005 : Klotz & Radić

The 1990's was the opening up of Chile after 17 years of ruling under the military government, Augusto Pinochet (1973–1990). At that time, architecture was one out of many cultural expressions of a new vibe, one that turned into a norm by the time I was a student at Pontifical Catholic University of Chile, School of Architecture (PUC) in the early to late 2000's.

The so-called "golden generation"[1] of architects who graduated in the early 1990's from PUC were the spearheads of this new image of contemporary Chilean architecture, and this story was well represented in *a+u 06:07 Chile – Deep South*. Specifically, Alejandro Aravena, Mathias Klotz, Smiljan Radić, Cecilia Puga and Sebastián Irarrázaval; all of whom till today are the main representatives of its local production.

From this group, 2 figures seemed to stand out the most over the last 3 decades, providing 2 very contrasting ways to practice architecture in Chile: they are Mathias Klotz and Smiljan Radić.

Klotz certainly left the biggest mark in how architecture was done in Chile during the 1990's and mid 2000's. The appeal of his work lies in its pragmatism, with easy to understand schemes that provided him with a big base of followers and emulators. His approach could be easily traced back to his former associates – Felipe Assadi, etc. – and in his students at the Central University of Chile from the mid to late 1990's – like Ramón Coz Rosenfeld, Marco Polidura Alvarez and Iñaki Volante Negueruela, who are featured in this book (see pp. 146–153). Take Klotz's opera prima, the Tongoy House built in 1991 (see p. 15, above), for example, the straightforward layouts he used to pull off during the first 15 years of his career have been fully absorbed and integrated by the local architecture culture in ways that has now been completely normalized. It is something you do not notice anymore because it is everywhere.

On the other side of the spectrum, we have Smiljan Radić. Despite having the same age as Klotz, he took a longer time to be appreciated in both Chile and abroad. His approach could be described as the ultimate postmodern architect: everything you see is a citation to Radić's inner world of references, sometimes too hermetic to be understood. Coincidentally, this approach brings him closer to the slightly older Swiss architects such as Valerio Olgiati and Christian Kerez, with whom he shares a deep interest in the work of the late Japanese architect Kazuo Shinohara (Radić's Prismí House, see p. 52–63). Shinohara's strict control to the way his work was published echoed in that of Radić's. A mystifying aura comes from his reluctance to have a website and to explain his work too much; or when he does, they are in a non-formal or non-academic manner. Everything is done on his terms. He does not care about what others thought of him: what matters to him is the expression of oneself through one's work, viscerally. This radical personal attitude has come to be understood and published in several monographs, and this recognition is partly due to the fact that all of us – architects – see in Radić what we want to be, an aspiration: somebody who is able to do what he wants, regardless of clients, budget or site constraints – even regardless of the media.

1990—2005：克洛茨与拉迪奇

智利在奥古斯托·皮诺切特长达17年（1973-1990）的军政府统治下，当时的建筑与其他众多文化形式一样，带着新时代的气象。到了21世纪00年代，我在智利天主教大学（PUC）的建筑学院求学时，这种建筑文化已成为一种社会常态。

所谓"黄金一代"[①]即20世纪90年代初期从智利天主教大学毕业的建筑师们，他们是塑造当代智利建筑新形象的先锋人物，大致来龙去脉在《a+u 06:07 智利——深远的南国》专辑中有过详细阐述。具体来说，他们分别是亚历杭德罗·阿拉维纳、马蒂亚斯·克洛茨、史密里安·拉迪奇、塞西莉亚·普加和塞巴斯蒂安·伊拉拉萨瓦。而直到今天，他们仍代表着智利建筑。

其中，马蒂亚斯·克洛茨和史密里安·拉迪奇应该是智利近30年中最杰出的两位建筑师，他们提供了两种截然不同的建筑实践方式。

克洛茨无疑在20世纪90年代至21世纪00年代中期的智利建筑实践中留下了最深的印记。他的作品魅力在于实用性与易读性，这为他带来了大量的追随者和模仿者。在许多其他建筑师身上，都可以看到他的影响，例如费利佩·阿萨迪等他的前同事们，以及他在20世纪90年代中期至后期任教于智利天主教大学时的学生们，如拉蒙·柯兹·罗森菲尔德、马可·波里杜拉·阿尔瓦雷茨和伊纳基·博兰德·内格鲁埃拉，他们的作品也收录于本书之中（第146-153页）。以克洛茨的处女作，建造于1991年的通吉住宅（第15页，上部）为例，他前15年职业生涯中所使用的这种简单明快的建筑布局被当地建筑文化完全吸收融合，如今已经成为范式。它无处不在，已经成为一种常态。

史密里安·拉迪奇与克洛茨同龄，但他们的设计理念迥然不同。相较于克洛茨，拉迪奇在很久以后才获得智利以及国际上的认可。他采用所谓"极致的后现代主义"设计手法，即所观一切皆为内心世界的引述。有时这种手法过于与世隔绝而几乎不能被理解。巧合的是，这种方法与瓦勒里欧·奥尔加蒂和克里斯蒂安·克雷兹等稍稍年长的瑞士建筑师的设计方法接近，而且他们三位都对已故的日本建筑师筱原一男的作品深感兴趣（拉迪奇设计的棱镜屋，见52-63页）。筱原发表作品的途径与方式始终贯彻着自己的主张，这一点被拉迪奇效仿。拉迪奇神秘的气息来自于他对建立个人网站及对作品做过多解释的抗拒，而当他不得不介绍自己的作品时，会以随意、轻松的措辞进行阐述。一切都按照他自己的方式进行，他不在乎别人的看法。对他而言，如何通过作品来表达自己的真情实感才是最重要的。这种激进的个人态度渐渐开始被理解并发表于几本专著中。在某种程度上，这种认可是因为我们所有建筑师都在拉迪奇身上看到了自己的理想与抱负：成为一个随心的人。无论业主提出任何要求，无论预算和环境上有什么限制，甚至无论媒体如何干预，他都始终忠于自己。

简而言之，在智利，马蒂亚斯·克洛茨是可以效仿的建筑师，而史密里安·拉迪奇是人们憧憬的存在。但在过去的30年中，这种情况几乎已经被亚历杭德罗·阿拉维纳的出现打乱了。关于阿拉维纳，我将在接下来的文字中进一步阐述。

2005—2010：互联网

20世纪90年代的后半段，建筑博客（专门针对建筑的线上平台）开始激增，其中创立于美国加利福尼亚州洛杉矶的Archinect[②]是此类网站中历史最悠久的，现在仍然在更新。这些网站起初是关于建筑、设计和艺术等相关领域的

In short, Mathias Klotz is the architect you can emulate while Smiljan Radić is the one you hope to be in Chile. However, in the last 3 decades, this situation has been disrupted (but not entirely) by the emergence of Alejandro Aravena – who will be elaborated further in this essay.

2005–2010 : The Internet

Architecture weblogs (online sites dedicated to architecture) started to proliferate during the last half of the 1990's, with *Archinect*[2] – managed from Los Angeles, California – being the oldest of them, still accessible. These sites started as personal logs of individuals interested in architecture and related fields, such as design and art, with some of them taking on a more professional or commercial approach like *designboom*[3] – founded 1999 in Milan.

In 2005, *ArchDaily*'s[4] prototype website in Spanish – *Plataforma Arquitectura* – was incubated in the patios of PUC. On those days when a+u was editing its issue a+u 06:07 which celebrated Chile's 15 years of notorious local architecture production, the country was leading its way – for better or worse – towards the digitalization of architectural media and culture at large. For almost 10 years, 2009–2018, ArchDaily was the world's most visited architecture website.

ArchDaily hastened the way architecture was published. It reshaped the celebrity phenomenon in our field and the role models for students in those years. From Koolhas to Ingels, from Moneo to Aravena: those willing to play the fast game of internet architecture in its early years are today's (2020) most visible figures.

2 of the seminal works of Chilean architecture in the first half of the 2000's did not make it to a+u 06:07 – Quinta Monroy in Iquique by Elemental[5] (2003, a+u 11:02) (see p. 16, top) and Poli House in Coliumo by Chilean-Argentinian practice Pezo von Ellrichshausen (2005, a+u 13:06). They only premiered in 2007 on Plataforma Arquitectura and later, on ArchDaily through the lens of Chilean photographer, Cristóbal Palma. This was instrumental to the success of this webpage company. Like most open-access websites, its revenue comes from digital advertising. As a consequence, a high volume of traffic is required to sustain them. To keep their followers coming back for more, they have to use what they had on hand - naturally, Chilean architecture production - as the main source of fuel for its media machine to generate enough fresh content. With specific metrics revealing which architects are more popular than others, these practices showed up more often while the rest are taken away from the spotlight. Alejandro Aravena (a partner in Elemental) and Pezo von Ellrichshausen clearly survived this bid, but most local practices – like cheap fuel – did not live up to ArchDaily's expectations. Nonetheless, in this issue, some of them - OWAR Architects, Max Nuñez, Emilio Marin, and Juan Carlos López - are introduced, and they will (finally) get to seize their opportunity back onto the international stage.

Yet, the result of being overexposed by ArchDaily meant that Chilean architecture would have to reinvent itself in the following decade.

2010–2015 : Beyond Building

Because of the super-cycle of high commodity prices between 2000 and 2014 (our main export good is raw copper), Chile was not particularly affected by the 2008 World Economic Crisis. We had enough fiscal reserves to overcome that particular storm.

After the 2008 crash, architecture here did not have a turn against the so-called "starchitecture", neither did we lean towards austerity like in Spain or Japan - as portrayed in the countries' national pavilions for the Venice Architecture Biennale 2016 curated by Alejandro Aravena. Precarious living conditions have always been the norm in Chile and Latin America for centuries now, and by 2011, it was too hard to play dumb.

The generation of the Penguin Revolution in 2006 - a massive movement of high-school students demanding education reform - turned into one with the college students, when tensions arose again in 2011 under President Sebastián Piñera (2010-2014). This is so-called Chilean Winter. And, the same digital revolution that provided us with Facebook, YouTube and our own ArchDaily was - this time - instrumental to the country's political unrest.

个人日志，后来转型为更专业化、有广告植入的商业化形式，与1999年在米兰成立的designboom③相似。

2005年，ArchDaily④西班牙语版的雏形Plataforma Arquitectura在智利天主教大学的平台上诞生。也就是说，当时a+u编辑部在策划《a+u 06:07》月刊要收录介绍智利15年来闻名世界的本土建筑时，不论好或是不好，智利就已经在朝着建筑媒体和文化的整体数字化方向发展了。并且在2009年至2018年的近10年里，ArchDaily几乎一直是世界访问量最大的建筑网站。

ArchDaily加快了新项目的发布速度。它重塑了建筑领域的名人现象，影响了学生们的榜样选择。从库哈斯到英格斯，从莫内欧到阿拉维纳，那些愿意在建筑线上媒体时代伊始就加入高速网络游戏的玩家们是当今最引人注目的人物。

很可惜的是，完成于21世纪00年代前半段的智利两大建筑杰作，在当时未能收录在《a+u 06:07》中，它们分别是ELEMENTAL建筑师事务所⑤在伊基克设计的金塔蒙罗伊社会住宅（2003,《a+u 11:02》,见第16页，上）和由智利、阿根廷建筑师联合成立的事务所——佩索·冯·艾奇绍森建筑事务所——在科里欧玛设计的波里住宅（2005,《a+u 13:06》）。通过智利摄影师克里斯托瓦尔·帕尔马的镜头，这两个作品在2007年于线上媒体Plataforma Arquitectura，也就是之后的ArchDaily上初次亮相。这样的独家素材也促进了线上平台的成功。像其他大多数开放式网站一样，平台不得不仰仗数字广告的收入。因此，它们需要大量流量来维持网页的运营。为了保持用户的回流和扩展，平台不得不使用现有的素材作为媒体机器的主要燃料，来产生足够的新鲜内容。这些现有素材当然便是智利的建筑作品了。特定算法揭示

This page, above: Tongoy House (1991) by Mathias Klotz. Photo courtesy of Mathias Klotz. This page, below: Casa Habitación (1992-1997) by Smiljan Radic. Photo by Gonzalo Puga.

本页，上：马蒂亚斯·克洛茨设计的汤戈依之屋，1991年；本页，下：史密里安·拉迪奇设计的CASA之屋（1992–1997年）。

Claims for an educational reform then turned into an overall revision of the entire socio-economic and political apparatus governing Chile since the return of democracy in 1990. Architecture was also part of that purge. If the previous term (2005-2010) was a celebration of Chilean architects and their most remarkable pieces, the next cycle (2010-) was more about self-reflection and criticality. For instance, Emilio de la Cerda - former member of OWAR Architects (see pp. 172–179) - became the director of PUC School of Architecture in 2014 with a clear agenda towards public responsibility in architecture. One of her first actions was to change the director of PUC's own media: ARQMagazine - arguably Chile's most relevant printed publication on architecture since the mid 1990's. New editor Francisco Díaz shifted ARQ focus from remarkable pieces (e.g. the next Poli House) to relevant topics in our field and society at large (e.g. ecology, freedom, rules, commons, etc.). This way, the building was no longer in the center of debate: it turned into a means to refer to something else that is seemingly more urgent.

This was a very radical turn. Because up until that very moment, the building had always been in the center of our local architecture discourse which lasted for decades since the military strike of 1973 - when censorship banned activism in architecture and all other fields.⑥

One could say Chilean architecture of the 1960's and early 1970's is nowadays more popular with the younger professionals and students than the "golden generation". New players became more visible with activism back in the 2010's: Plan Común, an architecture practice which turned into a political display; Umwelt, an office that focuses on protocols, procedures and researches matching their interest in proving their thesis on buildings; Grupo Toma, an organic collaborative of architects who despise architecture with capital "A" (see p. 20, top); and República Portatil, a militant unit in Concepción that takes Pezo von Ellrichshausen's formal language into informal and temporary conditions. They all share seemingly horizontal collaboration structures with a taste for the banal and neglected aspects of the city, and hold a very critical position towards architects preceding them in Chile. They also have an international network of supporters because of that confrontation with the local scene.

One thing about activism in architecture is that it usually positions its focus outside of the field, like in politics and environmental issues. By doing so, they oppose to the more traditional approach towards architecture which favors the autonomy of the discipline. Their antagonists tend to stress the apparent "ugliness" of the architecture emerging from this group, which in some cases is deliberate or just a byproduct of disinterest in its craft.

However, if the banal is able to reproduce itself without the aid of architects, it exposes the key paradox underlying this set of practices: Is it really worth the effort to design something apparently "undesigned"? Have we learned from the mistakes of the architects in the 1960's and their fascination with the architecture without architects?

2015–2020 : The Aravena Effect
Now let's go back to Alejandro Aravena.

He won the most prestigious award in contemporary architecture - the Pritzker Prize - in 2016, the same year he curated the most relevant cultural event in the field, the Venice Architecture Biennale. It is interesting how the local media reacted to the prize, most times replacing Pritzker for the title "Nobel Prize of Architecture" in a bid to make the recognition more accessible to the general public. It was also an attempt by Aravena to make architecture more understandable and relevant to society at large - just like when he shifted the focus from elite, second homes (Klotz &

了哪些建筑师更受欢迎,基于此,他们的作品也更多地出现,而其他人则越来越得不到关注。亚历杭德罗·阿拉维纳(ELEMENTAL的合伙人)和佩索·冯·艾奇绍森显然在这样的竞争中幸存了下来,但大多数当地事务所如同廉价燃料一样,没能达到ArchDaily的标准。尽管如此,本书还是介绍了其中的一些事务所和建筑师——OWAR建筑师事务所、马克斯·努涅斯、埃米利奥·马林和胡安·卡洛斯·洛佩斯,我们希望这是一次反击的机会,让他们回到国际舞台。

不过,在ArchDaily的过度曝光下,智利的建筑界在接下来的十年中将不得不进行自我改革。

2010-2015:超越建筑

智利以粗铜出口为主。得益于2000年至2014年间价格大范围上涨的"超级时期",智利在2008年世界经济危机中并未受到明显影响。充足的财政储备帮助国家顺利地渡过了那场金融风暴。

2008年的环球股灾之后,智利的建筑界对所谓的"明星建筑"并没有产生反感,也没有像西班牙或日本那样有着极简倾向,这些都可以从亚历杭德罗·阿拉维纳策划的2016年威尼斯建筑双年展国家馆中看到。几个世纪以来,智利人民的生活条件一直捉襟见肘,这也是拉丁美洲人民的常态。到2011年,我们对这样的事实已经不能视而不见了。

2006年爆发的"企鹅革命"由当时的高中生发起,这是一次要求教育改革的大型学生运动。2011年,"企鹅革命"的那代人与当时的大学生们联合了起来,于塞巴斯蒂安·皮涅拉总统任中(2010-2014年)再次引发了紧张的局势。他们的大规模抗议活动被称为"智利之冬"。正是Facebook、YouTube、ArchDaily这些平台所代表的数字革命对这一次智利的政局动荡起了重要作用。

对教育制度的抗议活动最终变成了对整个社会经济和政治体系进行修正的要求。这对国家统治机构来说,是自1990年民主制度回归以来的首次考验。建筑也是修正运动的一部分。如果说2005年至2010年是讴歌智利建筑师与他们最杰出作品的时期,那么2010年之后更多的则是反思内省与自我批评。例如OWAR建筑师事务所(第172-179页)的前成员埃米利奥·德·拉·塞尔达在2014年成为智利天主教大学建筑学院的院长时,明确提出了重视建筑公共责任的方针。她首先更换了智利天主教大学发行的《ARQ》杂志的责任总编辑,该杂志可以说是20世纪90年代中期以来关于智利建筑内容最切题、最详实的印刷出版物。新任总编弗朗西斯科·迪亚兹将《ARQ》的关注点从名作主义(例如接下来将要介绍的波里住宅)转移到与建筑领域甚至是与整个社会相关的主题(例如生态、自由、法规和公共用地等)。这样一来,建筑便不再是辩论的中心,它变成了讨论看似更紧迫议题的一种媒介。

这是一个根本的转变。自1973年的军事政变起,审查机构监管建筑及所有其他领域中的激进主义数十年,因此当地建筑界的话语中心一直是建筑物本身。[6]

可以说,相比于"黄金一代",20世纪60年代和20世纪70年代初期的智利建筑更受当今青年建筑师和学生的青睐。21世纪10年代的改革运动使激进主义回归大众视野,新的参与者也愈加引人注目。比如说共同计划事务所,他们从建筑实践转变为政治思想的展示;乌姆韦特事务所致力于将关注领域的协议、程序和研究用作建筑的命题;托马集团是由一群建筑师组成的有机合作组织,他们反对广义上的建筑学(第20页,上);以及康塞普西翁区的一个激进组织"便携式共和国",他们将佩索·冯·艾奇绍森建筑事务所的形态语言带入了日常场景和临时建筑之中。他们都是看似横向的协作组织,关心城市中被忽视或被置之不理的方方面面,并且对之前的智利建筑师们持批判的态度。他们与局势的对抗还为他们赢得了大批国际网络上的支持者。

建筑界激进主义的焦点通常在建筑领域之外,如政治和环保问题。如此一来,他们便不能用传统的、支持学科自治性的建筑体系来进行设计。不认同这一群体的人通常强调由他们所设计的建筑的"丑陋"。在某些情况下这样的设计是有意而为之,或者有的只是对建筑工艺不感兴趣。

可是,如果平凡的事物能够在没有建筑师的情况下自我生产,这就揭露了这种实践之下的最大悖论——设计显然是"未经设计的"东西,这真的有意义吗?我们从1960年代的建筑师们犯下的错误,从他们对"没有建筑师的建筑"的迷恋中究竟吸取了什么样的教训?

2015-2020:阿拉维纳效应

现在,回到亚历杭德罗·阿拉维纳。

阿拉维纳在2016年赢得了当代建筑界最负盛名的普利兹克建筑奖。同年,他策划了建筑界最大的文化活

Radic) towards social housing (Quinta Monroy, see p. 16, top). And whether we like it or not, he left a strong mark in the generation after the student movement in 2011.

We all agree in Chile that "he made it", and to some of us Chileans, this is particularly remarkable for someone who was not born and raised in elite circles. But this incredible notoriety has its backlash, as more exposure equals more scrutiny which often gets confused with envy.

Just like his younger colleagues confronting the old guard after 2011, the reaction from architects of a purer disciplinary kin happened instantly after Aravena's boom. Most architects to his left and right mistakenly disregard him as a compromised practitioner - one that works for big businesses, and plays political games with different parties depending on the situation. By holding on to his position, Aravena has made this dirty realism his standard. Yet, this poses a similar paradox that was noted in the work of his younger peers: if the project loses all its attributes during the design and building process because the world is a dirty place (and that's the way it is), is it really worth the effort?

After his breakthrough and more strongly after 2016, Aravena - the centrist architect - had the power to polarize the local scene. Activists have started to become more active than him (e.g. Plan Común, Grupo Toma) and the discipline-based, autonomous architects have become more introverted than ever (e.g. Max Nuñez, Amunátegui Valdés, Beals Lyon, Guillermo Hevia G., Thiermann Cruz, etc.).

The introverted could be divided in 2 subgroups: the articulated and the intuitives. The articulated are represented by offices such as Amunátegui Valdés and Thiermann Cruz. In their work, historical citations are seen all over. Due to the international nature of both practices (with one partner based in Chile doing the dirty work, and other in the U.S. teaching, researching and writing), they seem to be aligned with relatively older contemporaries abroad such as Johnston Marklee, Christ & Gantenbein or Go Hasegawa, all of whom have ties with the academy and tend to design with consideration to the vast cultural repository of architecture history. Their works are usually labeled as Postmodernist; a deliberate remix of old buildings in which parts are recognized as a whole.

The referential crew, on the other hand, are the intuitives, represented by people such as Max Nuñez and Beals Lyon. They are more locally rooted than Amunátegu Valdés and Thiermann Cruz. Their references are less clear, almost blurry. The rhetoric behind their projects is less strict and more ad-hoc to sites or other more pragmatic concerns. Considering they are not full time academics, their works appear less articulate, looser, but more spontaneous.

But, indeed, there are those who also fall in between these 2 subgroups with a slightly compulsive use of historical architecture references, like Cristián Izquierdo, Emilio Marín and Juan Carlos López. The latter two, with a taste for more recent sources - like CarmodyGroarke's Studio East Dining (London, 2010), which is observed in their layout for the Desert Interpretation Center (see pp. 146-153).

Reflecting back, the same argument could be used to reunderstand some figures of the generation that graduated from PUC in the 1990's who tend to use citation in clear (articulate) or opaque (intuitive) ways. Guillermo Acuña, Sebastián Irarrázaval, Cecilia Puga and Smiljan Radic are examples of such figures and are clearly on the opposite side compared to the dirty realism of Alejandro Aravena.

2020- : CHILE #2

Back to social unrest - this is, after all, Latin America.

October 2019 marked a turning point in Chile's recent history after the return to democracy.

October 18th, 2019: 2 weeks of student protests against the sudden rise of public transport fees in Santiago, went out of control and turned into riots and fires all over Chile. Sebastián Piñera, now in his second presidency (2018-2022), declared state of emergency on October 19th. He gave control of public spaces to the military and set a curfew - a flashback to the ruling under the military government of Augusto Pinochet (1973-1990).

October 25th, 2019: Up to 1.2 million people protested peacefully in downtown Santiago. Piñera halted the curfew and state of emergency a few days later. There were all kinds of demands on the streets - for better pensions, health, education, water rights, gender equality, new constitution,

动:威尼斯建筑双年展。有趣的是,当地媒体大多将普利兹克建筑奖宣传为"建筑界的诺贝尔奖",以便公众能更容易理解该奖项的重要性。而让建筑更易于被理解,并与整个社会更加相关,也是阿拉维纳的努力方向。他工作重点以金塔蒙罗伊社会住宅(第16页,上)为轴,从面向精英的度假居所(克洛茨与拉迪奇)转向了社会住宅。2011年学生运动之后的一代人无论喜欢阿拉维纳与否,都被他深深地影响着。

智利人都同意"他成功了"。对于非精英圈子出生的人们来说,这尤其了不起。但是过大的名气往往伴随着代价,大量的曝光等于更多的审视,而其中常常混有嫉妒的目光。

就像2011年之后,挡在青年建筑师之前的保守派老卫兵一样,在阿拉维纳发迹之后,当时的传统主流建筑师们也很快站在了他的对立面。阿拉维纳周围的建筑师们误认为他是一位折衷的从业者,为大型企业工作,并根据情况与各种政党玩政治游戏。但阿拉维纳坚持了自己的立场,他将这种肮脏的现实主义变成了自己的标准。然而,这使阿拉维纳也遭遇了类似上述青年建筑师们面对的悖论:如果一个项目在设计和建造过程中因为现实的肮脏(世界就是如此)而失去了所有的特质,那是否还值得努力去做?

阿拉维纳于2016年在事业上取得重大突破,并表现地更加强劲之后,作为中立建筑师,他将智利建筑界一分为二。激进派开始比他更加活跃(例如共同计划事务所和托马集团),而以学术为基础、相信建筑自律性的保守派,则比以往任何时候都更加自守于建筑领域内(如马克斯·努涅斯、阿姆纳特及·巴尔德斯、比尔斯·里昂、吉列尔莫·赫维亚·G和蒂尔曼·克鲁兹等)。

内向保守的建筑师中还可以再细分为分析派和直觉派。分析派的代表有阿姆纳特及·巴尔德斯和蒂尔曼·克鲁兹的事务所。他们的作品特征是随处可见的历史参照。这两个事务所是国际化的(一位合伙人在智利解决项目工作,另一位则在美国从事教学、研究与写作)。与国外一些老牌事务所,如约翰斯顿·马克利、克里斯特与甘藤斑或长谷川豪类似,他们都与学院有联系,并且在设计中使用整个建筑历史文化的知识。分析派的作品将旧建筑进行混合、编辑并突出局部,所以通常会被贴上后现代主义的标签。

直觉派的代表们则是马克斯·努涅斯和比尔斯·里昂等人。他们比阿姆纳特及·巴尔德斯、蒂尔曼·克鲁兹更加本土化,与当地关系密切。他们对历史的引用并不明显,可以说几乎是模糊的。他们项目背后的叙事不像教科书那般规矩,并对场地或其他更加务实方面有着优先的考量。他们不是全职的学者,所以他们的作品或许不那么清晰,甚至还有些许松散,但更加有自发的力量。

当然,也有一些人处于这两个派别之间,例如克里斯蒂安·伊斯基尔多、埃米利奥·马林和胡安·卡洛斯·洛佩斯,他们对历史建筑的引用略有强制性。后两者也从近年的建筑汲取灵感,在他们共同设计的沙漠文化中心的布局中(第146-153页),可以感知卡莫迪·格罗克事务所设计的东演播室餐厅的影响(伦敦,2010年)。

同样的方法可以用来回顾并重新理解1990年代从智利天主教大学毕业的那一代人,他们也倾向于用清晰(分析式)或不透明(直觉式)的方式引用历史建筑。吉列尔莫·阿库尼亚、塞巴斯蒂安·伊拉拉萨瓦、塞西莉亚·普加和史密里安·拉迪奇等就明显站在了亚历杭德罗·阿拉维纳现实主义的对立面。

2020–:智利2.0

社会回到动荡时期,这里毕竟是拉丁美洲。

自重返民主化之后,智利于2019年10月再次迎来历史性的转折点。

2019年10月18日,为反对圣地亚哥突然上涨的公共交通费,为期2周的学生抗议活动失控,在智利各地引发了骚乱和火灾。正处于第二任期(2018-2022年)的塞巴斯蒂安·皮涅拉总统于10月19日宣布国家进入紧急状态。为维持治安,他将公共空间的控制权交给了军队并实行了宵禁,使智利仿佛回到奥古斯托·皮诺切特时期(1973-1990年)的军政府统治之下。

2019年10月25日,多达120万人聚集于圣地亚哥市中心,进行和平抗议。几天后,皮涅拉总统解除了宵禁和紧急状态。民众提出了各种各样的需求:养老金、医疗和教育保障制度的改善、水的供给权、男女平等、新宪法的制定等。人们要求解决智利——这个在所有经济合作与发展组织(OECD)中最不平等、差距最大的成员国之一——所面临的社会经济问题。这一系列的抗议活动仍在继续,甚至到了2020年1月,智利仍处于严重的政治危机之中。

必须说明的是,我们在本书以及《a+u 06:07》中所收

19

etc. People were demanding for solutions to the socio-economic problems faced in Chile, which has one of the largest inequality gap among the OECD countries. Consequences of these events are still unfolding and the country remains in a state of deep political crisis even in January 2020.

It must be said that the architecture we celebrate in this book - and previously in a+u 06:07 - still belongs to the work of people in a privileged position. Access to commissions, clients and media is quite restricted to the ones outside of the circle (PUC and its orbits). Most of the building environments in Chile do not share the same level of quality as the 18 projects featured in this book. In a way, this body of architecture also expresses the social, economic and spatial inequality that most Chileans would find insulting.

From a different point of view, looking back at the production of a+u 06:07 fifteen years later, we could say that the architects in the "golden generation" were imagining a future though their work: they portrayed a place in which we will live after our problems are resolved - an oasis in a prosperous and peaceful country.

So, how do we get to that point - Chile #2? It may be for the younger generations to bridge the gap between these 2 contrasting realities.

Part of this responsibility lies with the academy. Using a designbuild method, University of Talca, School of Architecture[7] has concocted a recipe to bring architecture to places where it is needed the most. The method works with limited budgets and uses a sensitive approach to design that responds to context, echoing the early production of the "golden generation". But,

p. 16: Quinta Monroy Social Housing by Elemental (2003). Photo by Cristóbal Palma. This page, above: Centro de Creación Infante 1415 by Grupo Toma (2014). Photo by Infante 1415. This page, middle: Nancagua Municipality Building by Beals Lyon Architects (2015). Photo by Felipe Fontecilla. This page, below: Diploma Project Pinohuacho Vista Point by Talca University Student Rodrigo Sheward (2006). Photo courtesy of Rodrigo Sheward.

第16页：ELEMENTAL建筑事务所设计的金塔蒙罗伊社会住宅（2003年）。本页，上：托马集团设计的幼儿发展中心（2014年）；本页，中：比尔斯·里昂建筑师事务所设计的南卡瓜市政府大楼（2015年）；本页，下：塔尔卡大学学生罗德里戈·谢沃德的毕业作品皮诺瓦乔展望角（2006年）。

录的建筑都属于特权阶级的作品。在智利,建筑师若身处智利天主教大学这个特定圈子之外,接收委托的途径、与客户和媒体接近的机会都十分有限。本书介绍的18个项目所展示的人居环境质量水平高于全国大部分建筑。在某种程度上,这些高质量的建筑本身表达了社会、经济和空间的不平等,容易触及国民的逆鳞。

十多年后的今天,从不同的角度回顾《a+u 06:07》,我们可以说"黄金一代"建筑师的作品是对美好未来的想象。他们描绘了一个繁荣与和平的国家,是解决当下所有问题后的绿洲。

那么,我们要如何实现那个绿洲:智利2.0呢?是否能跨越这两个截然不同现实之间的鸿沟,这个重任就落在青年一代建筑师的肩上了。

学院承担着这个重任的一部分。塔尔卡大学建筑学院[7]采用设计与建造结合的教育方法,力图将建筑带到最需要的地方。该方法在预算有限的情况下,采用灵活设计响应当地文脉,这效仿了"黄金一代"的早期作品。但是,塔尔卡大学地处智利第二大城市(塔尔卡),对首都(圣地亚哥)的影响很小,而且该大学关注的农村地区仅覆盖了智利10%的人口,因此这个方法的通用性十分有限。尽管如此,塔尔卡式建筑教育自2000年代中期以来愈加具有影响力。例如,最近重建的美洲大学建筑学院(教职员包括那些创建了塔尔卡建筑学院[8]的学者)一直在圣地亚哥的郊外地区测试这种设计与建造结合的方法。智利几乎所有的政治和社会动荡都在这里孕育。然而,到目前为止,真正的建筑从未在此出现过。希望依然存在。

their scope is limited due to its location being in a secondary city (Talca) with very little influence on its capital (Santiago), and its focus on rural areas which covers only 10% of the Chilean population. Nonetheless, this method has been quite influential in educating architects since the mid 2000's and is now more relevant than before. For instance, in the recent re-founding of the School of Architecture in University of the Americas (which includes a few members of the first groups of academics who helped shape Talca's School of Architecture[8]), the school has been testing this design-build method in the peripheries of Santiago where most of the political and social unrest is incubated, and where architecture, up until now, had never showed. There is hope.

References:

1. The term "golden generation" was first coined in 2015 by local publication Revista Capital. This article featured an interview to Fernando Pérez Oyarzún who is one of the most influential academics in PUC from the 1980's up until today (2020) and is regarded as one of the masters of the architects included in this informal group, together with Montserrat Palmer, Teodoro Fernández and Rodrigo Pérez de Arce.

2. www.archinect.com

3. www.designboom.com

4. www.archdaily.com

5. Elemental S.A. was founded in 2001 by architects Alejandro Aravena and Pablo Allard, together with engineer Andres Iacobelli. Its original focus was social housing projects, while Aravena used to design non-social commissions under his own name. Today all of his projects are channeled through Elemental, running this office together with his partners Gonzalo Arteaga, Diego Torres, Víctor Oddó and Juan Ignacio Cerda.

6. Chilean architecture of the 1970's and 1980's had very limited visibility outside our country, probably due to our bad international reputation during those decades. It is worth mentioning the work of late Cristián Boza - our foremost postmodern architect - and Christian de Groote (mentoring figure to Luis Izquierdo and Antonia Lehman, featured in this issue), whose huge houses did not shy to our exuberant landscapes. Baby boomers Izquierdo, Lehman and Cristián Undurraga started their practices in the early 1980's, with good doses of professional and economic success which has been sustained up until today (2021).

7. Founded in 1999 by architect, academic and writer Juan Román, the original focus of University of Talca, School of Architecture was to train students to operate in the rural areas of Chile's Central Valley - in-between Santiago to the north and Temuco to the south. This is due to the origin of most of the students coming to Talca - first generation in higher education programs coming from agrarian areas. It could also be explained because it was a part of Chile which was not often considered in the mainstream discourse of architecture up until the 1990's. With a strong focus on construction and building experimentation, students in Talca have to build a piece of architecture as their diploma project. This venture must consider design, construction, fundraising, documentation and everything involved in the professional practice of architects.

8. University of the Americas, School of Architecture has been led by director Juan Pablo Corvalán since 2016. Co-founder and member of pan-american collective of architects Supersudaca, Corvalán was part of the first generation of academics that Juan Román brought to Talca School, together with Gregorio Brugnoli - also part of the team at University of the Americas. Their platform Línea de Intervención Comunitaria (LIC, translated as community intervention program) compromises the entire university and, in the School of Architecture, is producing build work all over Santiago as a collaboration of academics, students and local communities.

注释：

1. "黄金一代"一词最早出现在2015年智利本地的出版物《资本杂志》。该刊采访了费尔南多·佩雷斯·奥亚尊，他是智利天主教大学自1980年代至2020年以来最有影响力的学者之一，与蒙特塞拉特·帕尔默、特奥多罗·费尔南德斯和罗德里戈·佩雷斯·阿尔塞一起被看作是孕育了"黄金一代"的大师之一。

2. www.archinect.com

3. www.designboom.com

4. www.archdaily.com

5. ELEMENTAL建筑师事务所由建筑师亚历杭德罗·阿拉维纳和巴勃罗·阿拉德以及工程师安德烈斯·雅科贝利于2001年创立。事务所最初的重点是社会住房项目，同时阿拉维纳以个人名义承接其他委托。今天，阿拉维纳所有的项目都通过ELEMENTAL建筑师事务所承接，和他一起运营事务所的合伙人有贡萨洛阿尔泰加、迭戈托雷斯、维克多奥多和胡安·伊格纳西奥·塞尔达。

6. 可能是由于这几十年中智利在国际上的声誉不佳，1970年代和1980年代的智利建筑在国际上的知名度非常有限。值得向国际上推广的是我们最著名的后现代建筑师，已故的克里斯蒂安·博扎和克里斯蒂安·德·格罗特（本书收录的路易斯·伊斯基尔多和安东尼娅·雷曼的导师）的作品，他们的巨型住宅与智利壮美的风景相比也毫不逊色。早期建筑师伊斯基尔多·雷曼和克里斯蒂安·恩杜拉加于1980年代初开始实践，在专业和经济上皆取得了可喜的成就，并且一直持续到了今天（2021年）。

7. 塔尔卡大学建筑学院由建筑师、学者和作家胡安·罗曼于1999年成立，最初旨在培训能于智利中央山谷的乡村地区（北至圣地亚哥，南至特木科之间）开展业务的人才。因为大多数参与高等教育的第一代学生都来自农业地区。另外，1990年代之前，这块区域都不在建筑学主流的考虑之中。塔尔卡大学非常注重建造和建设的经历，因此学生的毕业设计是完成一个实体建筑。在此过程中，学生必须考虑设计、建造、筹款、程序说明书以及建筑师在专业实践中将会涉及的所有事项。

8. 自2016年以来，美洲大学建筑学院一直由胡安·帕布罗·科瓦伦领导。科瓦伦是泛美洲建筑师团体"超级苏达卡"的共同创办人及会员，也是胡安·罗曼邀请到塔尔卡大学的第一批学者。同来的还有格雷戈里奥·布鲁尼奥利，他现在也是美洲大学教职工的一员。他们的平台社区干预计划（LIC）覆盖了整个大学，并且在建筑学院内与学者、学生和当地社区合作，在整个圣地亚哥市内进行建筑实践。

Diego Grass studied architecture at Pontifical Catholic University of Chile (PUC). He joined Izquierdo Lehmann Architects as an associate from 2009 to 2011. In 2012 he formed Plan Común with Felipe de Ferrari, leaving in 2016 to form GRASS+BATZ+ together with Thomas Batzenschlager, focusing on projects mainly in Chile collaborating with NGOs and private commissioners. He has done audiovisual surveys on architecture since 2006, with all his productions compiled at database OnArchitecture (www.onarchitecture.com). He teaches at PUC (2014–), Talca University (2019–) and Harvard University Graduate School of Design (2020–).

迭戈·格拉斯在智利天主教大学学习建筑。他于2009至2011年加入伊斯基尔多·莱曼建筑师事务所担任合伙人。2012年，他与费利佩·德·法拉利创建了共同计划事务所，并于2016年离开。之后与托马斯·巴岑施拉格一起创立了格拉斯+巴兹+，主要致力于智利国内的民间组织和私人的委托。自2006年以来，他就对建筑进行视听调查，所有的成果在汇编后于数据库OnArchitecture (www.onarchitecture.com) 上公开。他在智利天主教大学（2014–），塔尔卡大学（2019–）和哈佛大学设计研究生院（2020–）任教。

Selected Works in Chile (2000–2020)
项目精选（2000～2020年）

1. Desert Interpretation Center
Emilio Marín, Juan Carlos López
Ayquina, Calama, Antofagasta 2013-2015
(pp. 146–153)

2. Museum of the Atacama Desert
Coz, Polidura, Volante, Soto
Huanchaca Ruins National Monument, Antofagasta 1996-2009
(pp. 124–133)

3. Ghat House
Max Núñez
Cachagua, Zapallar, Valparaíso 2015
(pp. 74–83)

4. Retiro Chapel
Undurraga Devés Architects
Celle Larga, Valparaíso 2008-2009
(pp. 134–145)

5. Dinamarca 399
Joaquín Velasco Rubio
Cerro Panteón, Valparaíso 2013-2014
(pp. 200–211)

6. Ruca Dwellings
Undurraga Devés Architects
Huechuraba, Santiago 2009-2011
(pp. 166–171)

7. Housing Complex in Quinta Normal
OWAR Architects
Quinta Normal, Santiago 2010-2015
(pp. 172–179)

8. Gabriela Mistral Cultural Center
Cristián Fernández Architects, Lateral Architecture & Design,
Santiago Commune, Santiago 2008 -
(pp. 190–199)

9. Palacio Pereira
Cecilia Puga, Paula Velasco, Alberto Moletto
Santiago Commune, Santiago 2012-2020
(pp. 212–223)

10. Cruz del Sur Building
Izquierdo Lehmann Architects
Las Condes, Santiago 2006-2009
(pp. 180–189)

11. Glass House
Max Núñez
Pirque, Santiago 2018
(pp. 154–165)

12. House for the Poem of the Right Angle
Smiljan Radić
Vilches, San Clemente, Maule 2010-2012
(pp. 40–51)

13. Loba House
Pezo von Ellrichshausen
Coliumo, Tomé, Biobío 2016-2017
(pp. 64–73)

14. Prism House + Terrace Room
Smiljan Radić
Conguillío, Araucanía 2017-2018
(pp. 52–63)

15. Casa 2Y
Sebastian Irarrazaval
Colico, Cunco, Araucanía 2013-2016
(pp. 28–39)

16. Palafito del Mar Hotel
Ortuzar Gebauer Architects
Castro, Los Lagos 2011-2013
(pp. 100–109)

17. Isla Lebe
Guillermo Acuña
Caleta Rilán, Castro, Los Lagos 2016-2020
(pp. 84–99)

18. Hotel of the Wind - Tierra Patagonia Hotel
Cazú Zegers Architects, Sarmiento Lake, Torres del Paine,
Magallanes y de la Antártica Chilena 2005-2011
(pp. 110–123)

Sebastian Irarrazaval
Casa 2Y
Colico, Cunco, Araucanía 2013-2016

塞巴斯蒂安·伊拉拉萨瓦
2Y 之家
阿劳卡尼亚，昆科，科里科　2013-2016

pp. 28–29: Aerial view of Casa 2Y. This page, above: View of the house overlooking Colico Lake. This page, below: Approach to the entrance of the house on the east. All photos on pp. 28-39 by Felipe Diaz Contardo.

第 28-29 页：鸟瞰 2Y 之家。本页，上：越过建筑，眺望科里科湖；本页，下：通往东部入口的道路。

There are 2 main obje1ctives of the house. First, it integrates the forest in the daily experience of the user. Second, it receives as much sunlight as possible during the entire day. For that purpose, the program fits into a two-Y-letter diagram that creates not only double orientations for the sun entering at different times of the day but also exposes the inhabitant to views that surround him. In other words, the extremely extended perimeter of the house and its bifurcations (in opposition to a compact organization) potentiates the experience of being – not in front of, but – from within. More specifically, of being among the sun and the trees in a parietal relationship. Additionally, its shape also made it possible to fit among the existing trees without having to clear the site. In regards to the materialization of the project, timber was chosen not only because it is a local material that could anchor the house to its place through affiliations with the natural and the cultural, but also because the wooden structures (in opposition with other kinds of material ensembles) are naturally built using infinite linear elements that as a result, resemble a sense of infinity that is present in forests and sort of echoes with the exterior. In this respect, the structural and architectural plans were done simultaneously in a permanent conversation. In the end, it is hard to distinguish one from the other. With regard to its relationship with the ground, the abovementioned 2Y diagram negotiates the slope in different manners. Sometimes, it underlines and rests upon the ground and other times, it contrasts with the soil, stressing the artificiality of its architectonic body. With respect to the treatment of its outer skin, the objective was to create a constant reverberation with its mostly green surroundings – and that, as it is well known, can be achieved using the color, red.

Credits and Data
Project title: 2Y House
Client: Private
Location: Colico, Chile
Design: 2013
Completion: 2016
Architect: Sebastian Irarrazaval
Design Team: Macarena Burdiles, Alicia Argüeles
Project Team: Felipe Cardemil (structural engineer), Jorge Ibacache
(construction)
Project area: 350 m²
Project estimate: 350,000 Euros

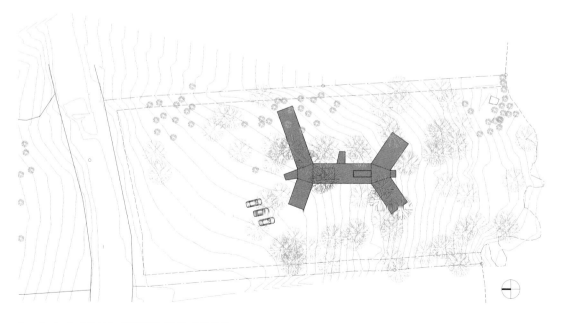

Site plan (scale: 1/1,000) ／总平面图（比例：1/1,000）

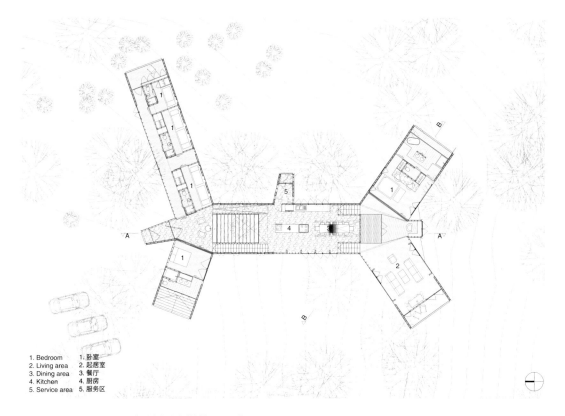

1. Bedroom　1. 卧室
2. Living area　2. 起居室
3. Dining area　3. 餐厅
4. Kitchen　4. 厨房
5. Service area　5. 服务区

First floor plan (scale: 1/400) ／一层平面图（比例：1/400）

建造这座住宅有两个主要目标。首先，它要能将森林融入住户的日常体验中。其次，它要能在一天之中充分地沐浴阳光。为了实现上述目标，这座建筑被设计成两个"Y"字的形状，这样不仅可以使一天中建筑受光的面积翻倍，还能让住户沉浸在周围的景色之中。换言之，这座建筑极力延伸的边缘及其分岔的造型与其他紧凑型的布局相反，强化了人们的体验——不是呈于眼前，而是身临其境；更具体地说，尝试建立了阳光和树林之间的有机媒介。此外，它的形状也使其可以适应现有的树木，而不必清理场地。在实际建造时，木材被作为最主要的材料使用，不仅因为它可以完全取材于当地，增加了建筑自然与文化的归属感；还因为木结构与其他类型的材料组合相反，是由无数线性的元素自然建造而成，代表了森林中存在的无限感，以及建筑与外部环境间的共振。在这方面，建筑的结构和平面仿佛在进行一次恒久的对话。最后，两者变得难以区分。就其与地面的关系而言，上述两个"Y"字的形状以不同的方式构筑在坡度之上。有时它强调地面本身；有时，它与土壤形成对比，强调建筑主体的人为性。关于建筑表皮的处理，为了与周围多为绿色的环境形成对比，使用了红色这一标志性的颜色。

This page: View from the west. The shape of the house aligns to the angle of its terrain.
本页：西侧外观。建筑依地形而建。

pp. 34–35: Exterior view from the terrace. This page: The kitchen and dining area extend onto an outdoor terrace. Opposite, above: The staircases connect the house both vertically and horizontally. Opposite, below: View of the living space on the lower "Y" section.

第 34-35 页，黄昏时的露台景观。本页：从厨房和餐厅，看向室外露台。对页，上：沿着斜面设置的楼梯。同时在水平和垂直方向上与住宅相连；对页，下："Y"字的节点低处的起居空间景观。

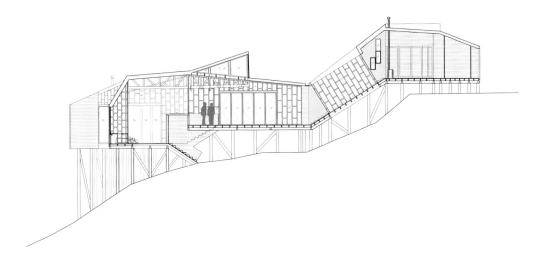

Section A ／剖面图 A

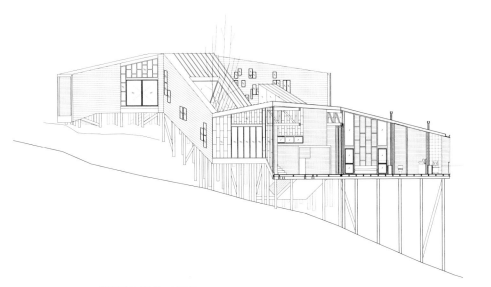

Section B (scale: 1/300) ／剖面图 B（比例：1/300）

Opposite: Interior view looking towards Colico Lake from between the "Y" intersection on the south end of the house.
对页：从建筑南端尽头"Y"字节点处的室内空间看向科里科湖。

Smiljan Radic
House for the Poem of the Right Angle
Vilches, San Clemente, Chile 2010-2012

史密里安·拉迪奇
《直角之诗》之家
毛莱，圣克利门蒂，比尔切斯，2010-2012

In the section on Flesh in Le Corbusier's The Poem of the Right Angle, the viewer occupies the body of a man lying in an ambiguous interior. He seems to be looking at an interior landscape where a woman is peering out towards an opening that reveals a passing cloud.

We can see the man's feet and a menhir, this area is covered by a hand placed domeshape on the whole lithograph, which increases the sense of an interior or a cavern. The cedar model of the House for the Poem of the Right Angle was first displayed in 2010 at the Global Endsexhibition at Toto Gallery-Ma in Tokyo. Its environment owes a lot to this picture and also to our installation in Venice, The Boy Hidden in a Fish – an opaque reinforced concrete exterior; a quiet, fragrant cedar interior . This house is the latest refuge that we have built. Its structure is a regular 12 cm thick reinforced concrete vault with a span of up to 15 m, resting on the perimeter walls. The vault is cut irregularly around its perimeter, with an open courtyard left inside. This blind volume faces a privileged mountain landscape, encrusted in an oak forest and hemmed in by an artificial garden containing 300 basalt stones. Its enclosure shows that its inhabitants are familiar with the surroundings – like the familiarity of a farmer, a hobo or a monk – naturally.

Credits and Data
Project title: House for the Poem of the Right Angle
Client: Private
Location: Vilches, Chile
Design: 2010
Completion: 2012
Architect: Smiljan Radić
Design Team: Jean Petitpas, Alejandro Lüer
Project Team: Marcela Correa (sculptor, landscape), Pedro Bartolomé (structural engineer)
Project area: 165 m² (built area), 4.5 ha (site area)

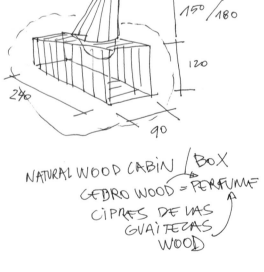

pp. 40–41: A scene of the house during winter. Photo courtesy of the architect. pp. 42–43: General view of the house from the east. Opposite, left: The Poem of the Right Angle, Le Corbusier. Image courtesy of F.L.C./ ADAGP, Paris & JASPAR, Tokyo, 2020 G2143. Opposite, right: The Boy Hidden in a Fish, Smiljan Radić. This page: 300 basalt stones are arranged into an artificial garden among the oak forest. Photos on pp. 40–45, p. 47, p. 51 (bottom) by Cristobal Palma.

第 40-41 页：冬天的景象。 第 42-43 页：东侧外观。对页，左：勒·柯布西耶《直角之诗》的插画；对页，右：史密里安·拉迪奇的"藏在鱼中的男孩"。本页：建在橡树林中、由 300 块玄武岩组成的人工花园。

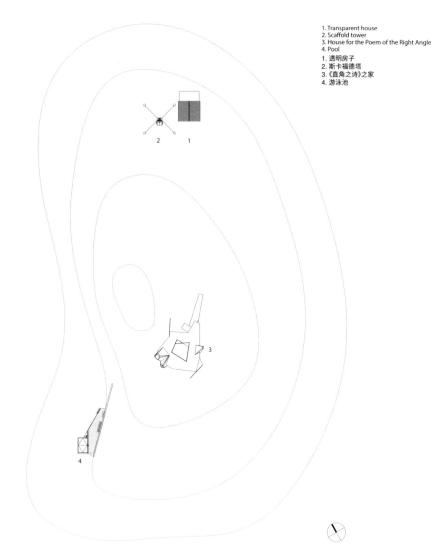

Site plan (scale: 1/1,500) ／总平面图（比例：1/1,500）

在勒·柯布西耶《直角之诗》中"肉体"一章的描绘里，占据画面主体的是躺在模糊的室内空间中的男性躯体。男人仿佛在观察室内的景观，而女人仿佛正通过一个洞口眺望外面天空中的流云。

我们能看到的是男人的脚和一块直立的巨石。整幅画面被一个手掌形状的穹顶所覆盖，强调了一种室内或是洞穴内的空间感。用雪松木建造的《直角之诗》之家模型首次展出于2010年东京TOTO展览中的"Global Ends exhibition"单元。它的环境借鉴了这幅画中的元素，以及我们在威尼斯的作品——"藏在鱼中的男孩"，一个不透明的钢筋混凝土外壳和一个安静、芳香的雪松木室内空间的结合体。《直角之诗》之家是我们最新设计建造的一座庇护所。其结构是一个常规的12厘米厚钢筋混凝土拱顶，跨度最大达15米，搭建在建筑外墙上。建筑体量沿着边缘做了不规则的切割，并在内部留有一个开放的庭院。这个封闭的体量正对着一片专属于它的山景，周围则被一片橡树林和一座有着300块玄武岩的人工花园包围。这种围合感象征着住户对周边自然环境的熟悉与亲近——就像农夫、流浪者或僧人一样。

This page: Exterior view of a truncated cone that brings light into the concrete vault.
本页:锥形的混凝土拱顶将光线引入室内。

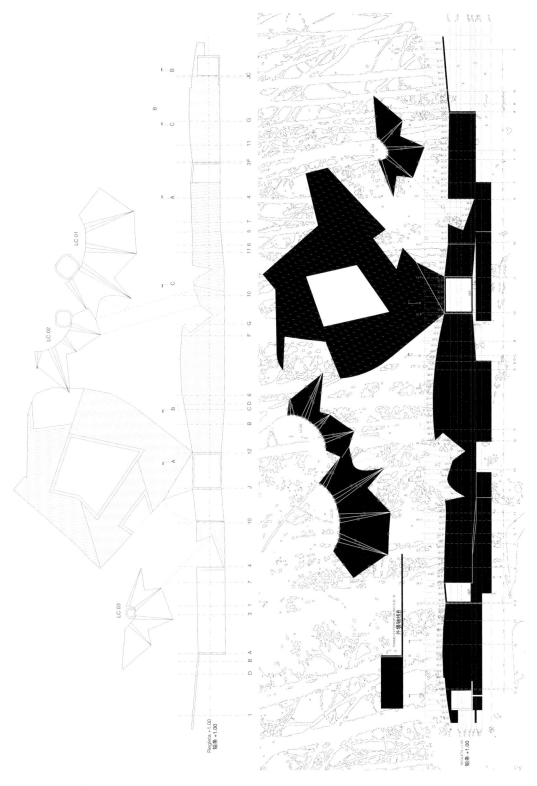

Unfolded elevation of the interior shell (top) and exterior shell (bottom) /展开的建筑表壳内侧立面和外侧立面

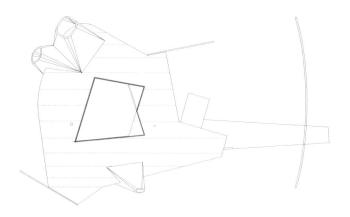

Roof plan／屋顶平面图

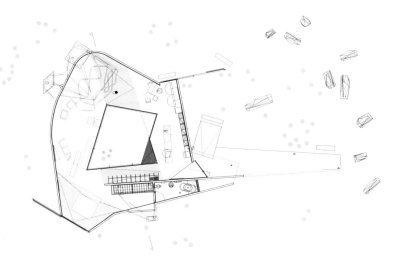

First floor plan (scale: 1/400)／一层平面图（比例：1/400）

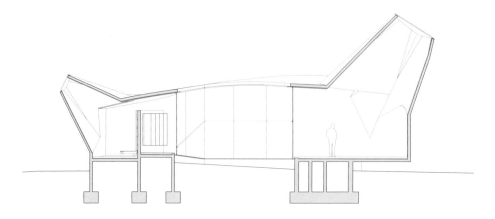

Section (scale: 1/200)／剖面图（比例：1/200）

p. 48: View from the internal open courtyard. This page, above: Interior view of the truncated cones. The hanging root-like sculpture appears to grow from the trees above, giving this feeling of nature extending into the vault. This page, below: View towards the fireplace in the living area of the house. Photos on p. 48, p. 51 (top) by Gonzalo Puga unless otherwise specified.

第 48 页：开放的中庭景观。本页，上：锥型空间内景，缠绕着的树根状雕塑仿佛生长在上空的树干上一样，给人以自然延伸到建筑内部的感觉；本页，下：带有暖炉的起居空间内景。

Prism House
Minamitsuru, Yamanashi, 1974

View of the northwest elevation / right: View of the northwest side wall from the bedroom

A sloped terrain passes freely underneath the terrace from where you can see the dead river of lava, a ghost recalling the last eruption of the Llaima Volcano. The Prism House by Kazuo Shinohara in 1974 stands on a similar platform (or what is left of it), as does the Room built in Chiloé in 1997 (or what is left of it).

I have attempted to replicate the air of the Prism House based on its geometrical structure. Its famous lateral facade, a right-angled isosceles triangle laid on the floor, and its exquisite interior diagonal wooden post, which like many other supports in the houses by this Japanese architect, orders the space by interrupting in a position that seems out of place, are signs of its "uncomfortable geometry". The exclusion of photographs of the longitudinal facade and of the bedrooms from publications leads me to believe that for the architect, the 45° cross-section of the volume and the post were the essence of the Prism House. I copied the structural cross-section of the original house, but I reduced its length from 10.80 m to 7.20 m. Thus, the Prism House facing the Llaima Volcano that we are building is effectively a section of a cube, approximating the ideal of Shinohara, assumed to be expressed in his photographs, and moving away from the reality of the Prism House in Japan, as it is simply a construction of a photographic piece of the original. From the Room, I replicated the last extension done to it. A galvanised steel structure supporting a red tent in the Chiloé forest. We applied a cross-section of a 7.20 m isosceles triangle to correctly accompany the size of its neighbor. We replicated its informal air and idea of creating only a large dormitory. To interpret is to pretend to do something else with the same, and that is of no interest to me. Thus, these 2 prefabricated prisms did not constitute an exercise in interpretation. In truth, this house is an exercise in repetition and replication, it is doing something again, though the gods may anger and the attempt always fails.

Credits and Data
Project title: Prism House + Terrace Room
Client: Private
Location: Conguillío, Chile
Design: 2017
Completion: 2018
Architect: Smiljan Radić
Design Team: Cristian Fuhrohp
Project area: 184 m² (built area), 0.6 ha (site area)

pp. 52–53: Aerial view of the Prism House in the forest of Conguillío National Park, overlooking a dead river of lava. pp. 54–55: General view of the Prism House (background) and Terrace Room (foreground). Opposite, above: Reference image. Room built in Chiloé by Smiljan Radić, 1997. Opposite, below: Reference image from JA 93. Prism House by Kazuo Shinohara, 1974. All photos on pp. 52–63 by Cristobal Palma.

第52-53页：鸟瞰位于孔吉利奥国家公园的棱镜屋，远处是岩浆凝固后形成的地表。第54-55页：直角三角屋（近处）和露台屋（远处）的概览。对页，上：史密里安·拉迪奇设计的房间（1997年）参考图；对页，下：日本杂志JA 93期提供的参考图。

倾斜的地形在露台下方自由奔跑，人们可以从那里看到熔岩的死河，幽灵般地提醒着伊拉玛火山的最后一次喷发。1974年由筱原一男设计的棱镜屋，以及1997年在奇洛埃建造的房间都坐落类似的平台上。

我尝试复制直角三角屋由几何构造创造出的氛围。它著名的内部横向立面，设置在地面上的等腰直角三角形，以及经常出现在这位日本建筑师其他作品中的精妙的室内斜向木柱，都是通过与周围环境格格不入的隔断，建立起空间的秩序，是筱原"不适几何学"的标志。然而，即便是有关这座建筑的出版物，也删除了有关纵向的立面以及卧室的照片。这更让我相信对于建筑师本人来说，45°的剖面以及柱子才是棱镜屋的精髓所在。我模仿了原建筑的剖面结构，但将它的长度从10.8米缩减到了7.2米。因此，我们这座面对伊拉玛火山的棱镜屋，实际上是一个立方体的一部分，它更接近筱原一男理想中的状态，仿佛是将原版建筑的照片转换成了精简的建筑语言，并在现实感上与日本的原作拉开了距离。在1997年建造的露台屋项目中，我在对它的最后一次增建中再次尝试了对棱镜屋的重现，而落成的就是奇洛埃森林里一个镀锌钢结构支撑着的红色帐篷。我们用了一个7.2米等腰三角形的横截面来准确地适配它相邻建筑的大小，重现了它非正式的氛围和只设计一个大卧室的想法。所谓诠释，就是假装把同样的东西当成新的东西来对待，对此，我兴味索然。因此，这两个预制结构组成的棱镜并非诠释的实践，而是一个有关重复和复制的实践。尽管这种尝试总是招来神之愤怒和失败的结果。

Opposite: West elevational view of the house during winter.
对页：西侧立面，冬季的景象。

Opposite: View of the bedroom on the upper floor of the Terrace Room with an intimate atmosphere. This page: In contrast, the bedroom in the Prism House is more open, with a full view of the forest.

对页：露台屋上层的卧室景观，有一种亲切的氛围。本页：相比之下，棱镜屋的卧室更为开阔，可以看到森林全景。

This page, above: Interior view on the lower floor of the Terrace Room, facing the hillside. This page, below: Each of the 2 stairs lead to the bedrooms on the upper floor.

本页，上：面向山区，露台屋下层的室内景观；本页，下：两边的楼梯都通往楼上的卧室。

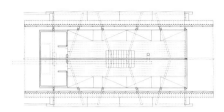

Second floor plan ／二层平面图

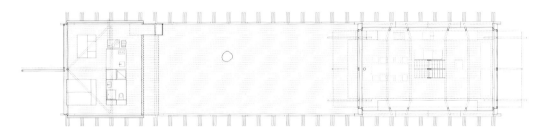

First floor plan (scale: 1/300) ／一层平面图（比例：1/300）

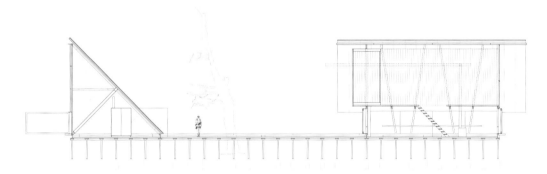

Long section (scale: 1/300) ／横向剖面图（比例：1/300）

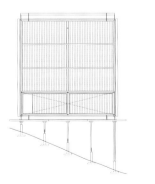

East elevation (scale: 1/300) ／东立面图（比例：1/300）

West elevation ／西立面图

Pezo von Ellrichshausen
Loba House
Coliumo, Tomé, Biobío 2016-2017

佩索·冯·艾奇绍森
罗巴住宅
比奥博,托默,克琉默 2016-2017

Site plan (scale: 1/5,000) ／总平面图（比例：1/5,000）

Perhaps the only distinction between objects and things resides in their scale. Closer to any natural thing, in its ambiguous scale, this small building is more than a hut but less than a house: it is a cottage. As an opaque block, a monolithic object heavily anchored at the edge of a cliff, it is facing a sea-lion reserve on the Pacific Ocean. In its under dimensioned thickness, in its narrow and tall proportion, the building could be read as an inhabited wall that runs perpendicular to the natural topography. The height of this wall is determined by 2 lines: a continuous horizon and a stepped sequence of 6 platforms that descend towards the sea. The separation between that horizontal roof (with the function of an open terrace) and the regular extension of the ground (with the informal arrangement of rest, dining and living), is a single asymmetrical room interrupted by 3 massive columns and 2 bridges. While beds are placed in the upper platforms with low ceiling, sofas or tables are meant to be in the lower platforms within a vertical space. There is a discreet regime of openings on either sides of the long volume with some punctual skylights, a few half-moon perforations that could be used as sun clocks and a singular corner window divided by a round pillar. This is the only window with unframed glass flushed to the outer concrete surface. Mirroring the sunset, an almost impossible and illusory floating rock rests right on top of that reflection.

pp. 64–65: General view of the house anchored onto the edge of the cliff. p. 66: View from the roof looking out onto the vast Pacific Ocean. All photos on pp. 64–73 courtesy of the architects. This page, above, both images: Concept paintings by the architect. This page, below, both images: Physical model of the house revealing its openings and 6 stepped platforms.

第 64-65 页：从东侧看，镶嵌在山崖端处的住宅。第 66 页：从屋顶远眺广阔的太平洋。本页，上两图：概念草图；本页，下两图：表现建筑开口位置和内部 6 个阶梯状平台的实体模型。

也许物件和事物之间唯一的区别在于它们的尺度。这座小建筑的尺度颇为暧昧，比小屋略大，却比住宅略小，这使它更接近于自然事物。它如同一个不透明的体块，紧紧地被固定在悬崖边缘，面朝着太平洋上的一个海狮保护区。考虑到它远低于平常的宽度，以及瘦高的比例，这座建筑可以被理解为一面为了居住而设立的墙，直插地表之上。墙的高度，由连续的地平线，与6个平台形成阶梯状、面朝大海的两条线条来界定。水平屋顶（具有开放露台的功能）和地面的常规扩展空间（休息、用餐、生活随性的布置），由3根粗壮的柱子和2座桥分割开来，形成一个非对称的房间。在一个垂直的空间内，床被放置在天花板较低的上层平台上，而沙发或桌子被放置在较低的平台上。建筑纵向两侧都设置了低调的开口，几面天窗被一丝不苟地打开，半月形的洞口可被用作太阳钟，还有一个被圆柱分割的别致角窗。这是建筑内部唯一的无框窗户，装有无框玻璃并和混凝土外墙平齐。映照着夕阳，整个建筑仿佛一块不真实的、虚幻的浮石坐落在倒影之上。

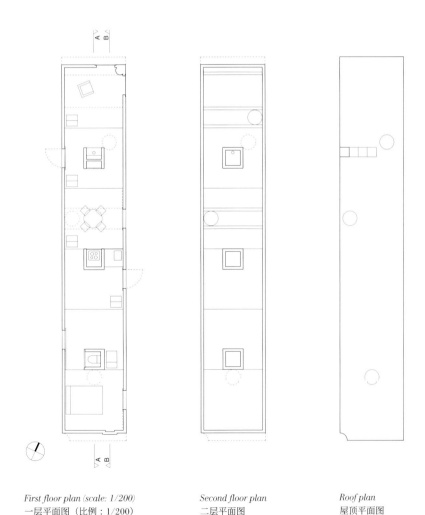

First floor plan (scale: 1/200)
一层平面图（比例：1/200）

Second floor plan
二层平面图

Roof plan
屋顶平面图

This page: View from the 5th platform facing the 6th, a proposed space for a bed. Opposite: View from the highest platform looking down across the space below.

本页：从第五段阶梯看向第六段阶梯——设想中作为卧室的场所。对页：从最高的阶梯上俯视整个室内空间。

Credits and Data
Project title: Loba House
Client: Marcelo Sánchez, Janis Hananias
Location: Coliumo, Tomé, VIII Region, Chile
Design: 2016
Completion: 2017
Architects: Mauricio Pezo, Sofía von Ellrichshausen
Design Team: Diego Perez, Thomas Sommerauer, Teresa Freire,
    Beatrice Pedroti,
Wiktor Gago
Project Team: Carvajal & Cabrer (builder), Peter Dechent
    (structure), Marcelo
Valenzuela, Daniel Garrido (consultants)
Project area: 70 m² (built area), 1,000 m² (site area)

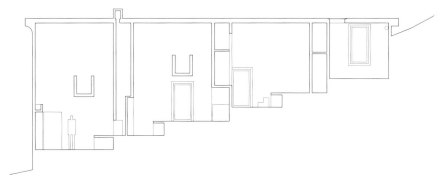

Section A /剖面图 A

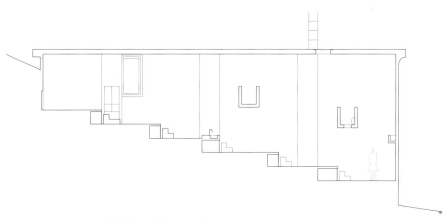

Section B (scale: 1/200) /剖面图 B（比例：1/200）

Opposite: The corner windows, divided by a round pillar, are covered by an unframed glass that is flushed to its outer concrete surface.

对页：被圆柱分割的角窗，覆盖着与外表面平齐的无框玻璃。

Max Núñez
Ghat House
Cachagua, Zapallar, Valparaíso 2015

马克斯·努涅斯
盖特住宅
瓦尔帕拉索，扎帕拉，卡查瓜 2015

The house is located over a terrain with a 25° incline facing the Pacific Ocean. Its design, structure, internal organization and lifestyle within it are determined by the site topography and the decision of creating a continuous space over the preexisting slope. The inclined reinforced concrete slab of the roof is parallel to the natural gradient of the site.

The roof is occupied by a large outdoor living area of various levels which are part stairs and part terraces with unusual proportions for a domestic program. Below this oblique plane, a continuous diagonal interior space contains the different public programs of the house. The monotony of its free plan is redefined by the slope, creating an interior topography of 11 levels and 7 stairs. The roof is supported by 15 concrete columns with different sizes and shapes. The irregular geometry that characterized them is determined by their structural needs and spatial aspects. Their heterogeneous shapes succeed in individualizing each column as a singular element, thus avoiding the reading of a rigid structural grid dictating its plan. Each column generates a particular point in space and the framing of the landscape between them is always diverse. 4 lighter volumes cladded in lenga wood interfere with the surface of the roof and the space below it. 3 of these volumes contain private rooms. The fourth, a smaller volume, allows direct access to the terrace from the interior. These volumes are located under, beside and above the roof establishing ambiguous relations between the private and public areas of the house.

Credits and Data
Project title: Ghat House
Client: Private
Location: Cachagua, Zapallar, Valparaíso, Chile
Completion: December 2015
Architect: Max Núñez
Design Team: Stefano Rolla
Project Team: Mauricio Ahumada (structural engineer), Francisco Álvarez (building contractor), Alejandra Marambio (landscape), Estudio Par (lighting design), Alfonso Bravo (technical inspection)
Project area: 390 m²

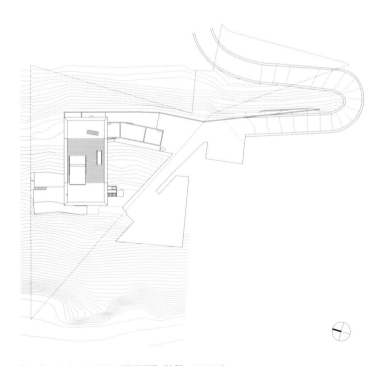

Site plan (scale: 1/1,200) ／总平面图（比例：1/1,200）

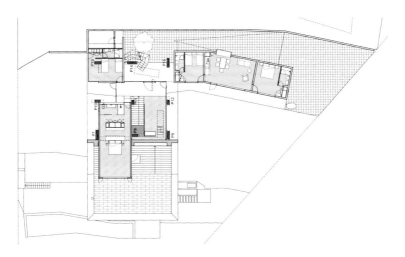

Second floor plan ／二层平面图

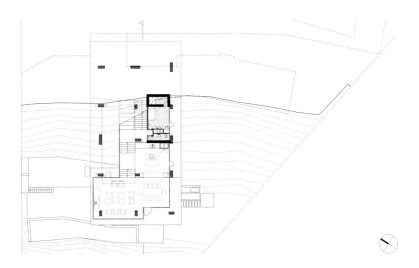

First floor plan (scale: 1/500) ／一层平面图（比例：1/500）

pp. 74–75: View from the roof of the house, overlooking the Pacific Ocean. pp. 76–77: The house's design, structure and organization follows its site topography to create a horizontal continuous space along the slope. All photos on pp. 74–83 by Roland Halbe.

第 74-75 页：从屋顶远眺太平洋。第 76-77 页：住宅的设计、结构和布局均遵循基地地形变化，从而创造了沿斜坡布局的一系列连续水平空间。

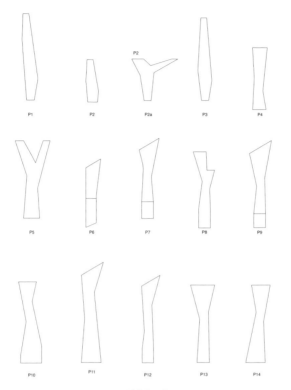

Structural column diagram /结构柱示意图

住宅坐落于面对太平洋的 25°倾斜的坡道上。场地地势以及在当地建造一个连续空间的想法，决定了这座建筑的设计、结构、内部空间和生活方式。倾斜的钢筋混凝土顶棚与场地的坡度互相平行。

屋顶上设有多种由台阶和露台组成的室外生活空间，这种构成在智利国内项目中非常罕见。在倾斜平台之下，是一个连续的斜向室内空间，包含了多样的公共功能。斜坡在室内空间中衍生出 11 个层次和 7 段楼梯，大大降低了单一空间带来的乏味感。屋顶的支撑结构由 15 根形状尺寸各异的混凝土柱组成。它们独特而富有个性的形状，取决于各自的结构需要和空间要素。丰富多样的形状，成功地让每根柱子成为一个独立的元素，并且避免了规整的柱网所带给人们的刻板印象。每根柱子在空间中作为一个特殊的点存在，它们之间的景观框架各不相同。4 个由莲茄木包裹的轻型体块穿插在屋顶和屋顶下的空间内。其中 3 个是私人房间，第 4 个更小的体块则直接作为连通屋顶露台和室内空间的通道。这些体块被设置在屋顶的上下左右，这样的设计使这栋建筑内的私人空间和公共区域间的界线变得更加暧昧。

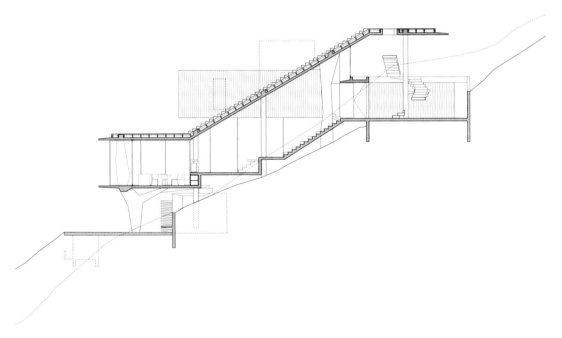

Section (scale: 1/250) ／剖面图（比例：1/250）

p. 81: The roof is supported by 15 concrete columns, each having a unique shape and size. Opposite: Interior view from the kitchen. The programs of the house are staggered onto different levels. This page: The living and dining room of the house faces a wide unobstructed view of the ocean.

第81页：屋顶由15根混凝土柱支撑，每根柱子都有独特的形状和尺寸。对页：厨房内景。房子的各个功能被配置在不同的楼层上。本页：起居室和餐厅面朝一片没有遮挡的海景。

Guillermo Acuña
Isla Lebe
Caleta Rilán, Castro, Los Lagos 2016–2020

吉列尔莫·阿库尼亚
莱贝岛
湖区，卡斯特罗，卡莱塔里尔 2016-2020

The project is located on one of the extraordinary intercoastal islands of the Chiloé archipelago, around 150 m from the coast of the southern bay of Rilan. Spanning 600 m in length, 80 m in width, and with an orientation towards 100º East, the small 5-hectare island looks towards the fjords and channels that make up the Desertores archipelago, between the Michinmahuida and Corcovado volcanoes. Its inter-tidal condition provokes a constant change in the landscape. Every 6 hours, large masses of water flood then empty the horizon, connecting and disconnecting the island from its nearby coast as the marine ground appears and disappears almost without warning. The island floats within this landscape of continuous change in which the natural elements – air, water, fire and earth – seem to be animated by a higher force. It is here that a series of small constructions are inserted. Simple and austere, they experience the restlessness of the environment, so that not much else is needed to inhabit this place. Big enough to provide shelter and small enough to induce us back to the sea, four 60 m² constructions and a series of walkways similar to those of the town of Tortel constitute a cove. The constructions have been built over time and their uses and programs have changed. The first is a boathouse that became a cellar and red room in which to meet and eat. An exterior staircase connects to an upper floor with 2 guestrooms. A terrace and few walkways lead to 2 houses that are almost identical (one proportionally larger than the other), and continue through the reforested site before finally arriving on the beach. The constructions change as the landscape changes.

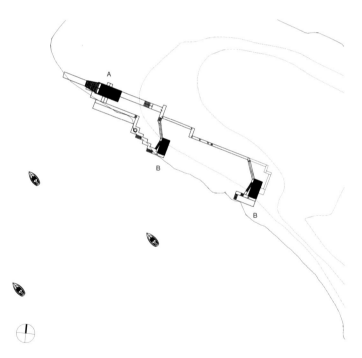

Site plan (scale: 1/2,000) ／总平面图（比例：1/2,000）

pp. 84–85: *Aerial view of the 3 houses sitting along the banks of an intercoastal island, Isla Lebe. Opposite: View of the boathouse with a large exterior stairs leading to the upper floor. All photos on pp. 84–99 by Cristobal Palma.*

第 84-85 页：鸟瞰建造在奇洛埃诸岛内侧莱贝岛海岸的 3 座建筑。对页：船屋建有通向上层的大型室外楼梯。

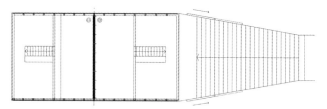

Loft floor plan of boathouse (A) /船屋（A）阁楼层平面图

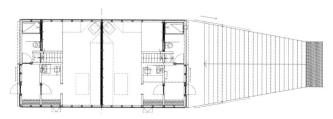

Second floor plan of boathouse (A) /船屋（A）二层平面图

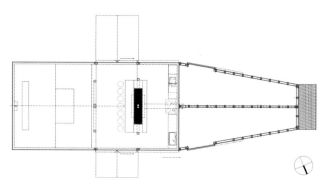

First floor plan of boathouse (A) (scale: 1/250) /船屋（A）一层平面图（比例：1/250）

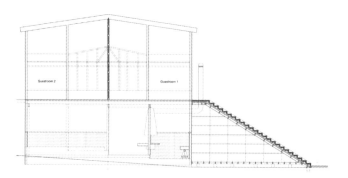

Long section of boathouse (A) (scale: 1/250) /船屋（A）横向剖面图（比例：1/250）

p. 89: A boardwalk on the upper floor connects the boathouse to the other 2 houses. This page: Exterior view along the large steps of the boathouse. Opposite: Main entrance of the boathouse on the lower floor.

第 89 页：上层的木板路连接起了船屋和另外两栋房子。本页：船屋外巨大楼梯的景观。对页：船屋下层的主入口。

Opposite: Interior view of the red room in the boathouse where the family or guests meet and eat. This page: Guestroom on the second floor of the boathouse.

对页:船屋内的红色房间是家人会客和吃饭的场所。本页:船屋二楼的会客室。

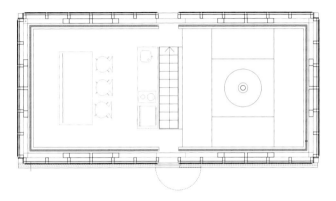

Second floor plan of the house (B) ／住宅（B）二层平面图

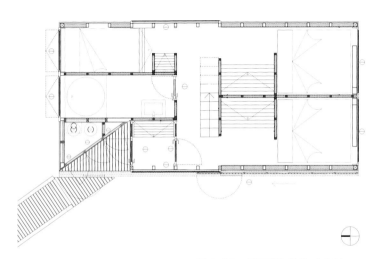

First floor plan of the house (B) (scale: 1/100) ／住宅（B）一层平面图（比例：1/100）

这个项目坐落于距离利兰南湾海岸 150 米，奇洛埃群岛南岸的一座小岛上。这座小岛长 600 米、宽 80 米，面朝偏东方 100°，面积为 5 万平方米，与位于明钦马维达和科尔科瓦多火山之间的迪斯托瑞斯群岛中的峡湾杣运河隔海相望。潮汐带的环境对地形产生了持续的影响。每 6 个小时，大量的海水涌入，几乎将地平线淹没，切断了小岛和近处海岸的联系，也使近海的地面忽隐忽现。这座小岛漂浮于不断变化的自然环境之中，空气、水、火、土，仿佛被注入了旺盛的生命力。就在这样的环境下，一系列小建筑被嵌入。不安的栖居状态，使它们斫雕为朴，简单而没有任何不必要的装饰。它们的尺度大小足以作为避风所，同时却也小得能让我们充分感受到大海的气息。4 座 60 平方米的托尔特尔小镇风的建筑与一系列步道相连，形成一条连续的弯廊。建筑的建造经历了相当长的时间，它们的功能和秩序也发生了变化。第一个被改造的是一座船库，它现在变成了一个下层的红色房间，供人们聚餐。外部的楼梯连接着楼上的两间客房。露台和连廊与另外两座几乎一样的房子相连（一座比另一座略大），并且一路经过被加固过的建筑基地，最终通向海滩。所有的建筑结构都随着地形的变化作了相应的改变。

Credits and Data
Project title: Isla Lebe
Client: Guillermo Acuña, Roberto Pons
Location: Isla Lebe, Caleta Rilán, Castro, Los Lagos
Design: 2016
Completion: 2020
Architect: Guillermo Acuña
Design Team: Guillermo Acuña Associate Architects
Project Team: Luis Miranda (consultant)
Project area: 315 m² (built area), 50,000 m² (site area)

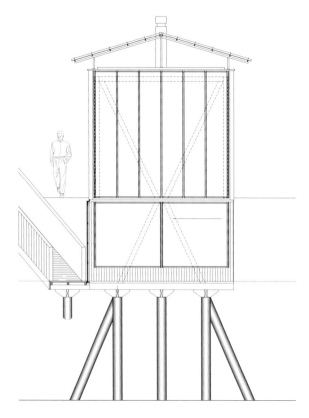

South elevation of the house (B) (scale: 1/100) ／住宅（B）南立面图（比例：1/100）

pp. 94-95: View of one of the 2 identical houses from the shore. The larger of the houses can be seen in the background. p. 96: Approach to the house, an intimate entrance hidden among the vegetation. Opposite, both images: Interior views of the house on the upper floor. This glass-covered house provides a view of the vegetation on one end, and an open sea view on the other.

第 94-95 页：海边的两座相同的住宅，远处的一座面积较大。第 96 页：通向小屋的道路和一个隐藏在植被中的秘密入口。对页，两图：住宅上层空间的内景。这座由玻璃覆盖的房子一端可以看到植被，另一端可以看到开阔的海景。

Ortuzar Gebauer Architects
Palafito del Mar Hotel
Castro, Los Lagos 2011–2013

奥尔图扎尔·格鲍尔建筑师事务所
帕拉非托德马尔酒店
湖区,卡斯特罗 2011-2013

Palafito del Mar Hotel is part of a 6 architectural projects for the recovery of a neighborhood Castro Chiloè island. The other 5 projects are Hostel Palafito Sur, Palacito Apart Hotel, Patio Palafito Hotel and Cafeteria, Fisherman's Place and Building Station.

The stilt houses (known as palafitos in Chile) on Pedro Montt street, one of the 4 palafitos neighborhoods in Castro city, was very depressed and in constant abandonment by its inhabitants who considered their sector to be marginal and poor. It is difficult to understand the condition in which this palafitos neighborhood was, despite being an image that is exported from the country as a heritage value. Since 2011, the interventions made are born from the observation of what the neighborhood needed – starting from our experience and choice to live in it, since, in addition to being architects, we are part of the community, as inhabitants and neighbors. The realization of the 6 projects, allowed many of these constructions on piles – that were abandoned or in deep state of deterioration – to be reused, recovered or reconverted.

This recovery also encouraged the "self-revitalization and revaluation" of its people by their neighborhood – to think about what has turned the neighborhood into a thriving tourist district in which people and visitors coexist, thus renewing its color and infrastructure.

Palafito del Mar Hotel is a design project with a fresh look on the original way of living in Chiloé – on stilts. The rooms of the hotel have patios and balconies overlooking the sea, and are linked by a central corridor that brings in light. Finally, a large room at the end functions as a communal square or meeting place around the fire. From there, the view of the austral canals gives meaning to the construction on stilts. The commission of the project was to design a boutique hotel as an experience to live in Chiloé, where all the bedrooms overlook the Castro estuary, and where the tides are always present in every corner of the project. Thus, the idea of making a hotel that represents a neighborhood of pile dwellings – a collage composition with different colors, shapes and textures specific to keeping the traditional granule that characterizes the neighborhood. This distribution will permit every room to have a seaview and a terrace while keeping its privacy. A central circulation enhanced by a skylight leads to the different rooms arranged linearly, successively crossing the various "thresholds of the sea" where one observes the tide from inside the stilts.

Finally, it concludes in a living room – by the fire – that welcomes guests for an encounter of Chiloè culture.

Credits and Data
Project title: Palafito del Mar Hotel
Client: Francisco Valdés y Sara Bertrand
Location: Castro, Chiloé, Chile
Design: 2011–2012
Completion: 2013
Architect: Eugenio Ortúzar, Tania Gebauer
Design Team: María Teresa de la Fuente
Project area: 300 m² (including terraces)

pp. 100–101: General view of the Palafito del Mar Hotel from across the river. Opposite: Interior view looking out onto the palafitos neighborhood along the banks of Rio La Chacra river. All photos on pp. 100–109 courtesy of the architects unless otherwise specified.

第 100-101 页：从帕拉非托德马尔酒店对岸看向酒店。对页：从酒店内部看里奥拉查克拉河沿岸的底层架空式建筑群。

Site map /项目区位示意图

This page, below 3 images: Historical photos of Castro Chiloë island. Opposite, 4 images, clockwise from top left: Hostel Palafito Sur. Palacito Apart Hotel. Patio Palafito Hotel and Cafeteria. Building Station. These projects are part of the recovery of a neighborhood in Castro Chiloë island.

本页，下三图：曾经的卡斯特罗市。对页，四图，从左上起顺时针：南帕拉非托酒店、帕拉非托公寓酒店、帕拉非托花园酒店和自助食堂、建筑中心。这些项目都是卡斯特罗奇洛埃群岛修复计划的一部分。

帕拉非托德马尔酒店是位于奇洛埃群岛卡斯特罗市内的一个旧城改造项目，同样的项目共有 6 个。另外 5 个分别是南帕拉非托酒店、帕拉非托公寓酒店、帕拉非托花园酒店和自助食堂、渔人之家以及建筑中心。

这是一个依靠浮柱支撑的架空式建筑群（在智利被称为"Palafitos"），位于佩德罗蒙特大街上。和它类似的架空式建筑群还有 3 个，但因当地居民认为其所处的地区贫穷且被边缘化，所以它们都经历了长年的萧条和废弃。除了被作为国家文化遗产来展示之外，很难再从其他角度来理解这些建筑的处境。从 2011 年开始，我们决定根据自身的经验和实际居住后的感想，既作为社区的一员，同时也以建筑师的身份，来观察、体验什么才是这里最需要的，并着手开始改造这些地区。这 6 个项目的实际建造过程中，许多已经严重劣化甚至被废弃的建筑得以重见天日，恢复，重新规划并被再次利用。

同时，这些项目倡导"当地人的自我振兴和自我价值观实现"，引领他们去思考如何将当地转变为一个旅游胜地，吸引大量游客。而最终得出的答案是更新当地的色调和基础设施。

帕拉非托德马尔酒店刷新了这种奇洛埃群岛传统的建筑形式——架空式建筑，在人们心中的印象。它的房间均带有能够观赏大海的露台和阳台，并由一个中央长廊所串联，将阳光带入室内。尽头的一个大房间则作为公共空间来使用，人们围绕着火炉交谈或者进行其他活动。在这里，南部运河的美景与建筑临水而居的环境相得益彰。这个项目的初衷是设计一座可以体验奇洛埃群岛风土人情的精品酒店，所有的卧室都能远眺卡斯特罗河口，让入住的客人时刻都有仿佛置身于室外的潮汐之中的体验。同时，酒店作为当地架空式建筑的象征，融合了不同的色彩、形式和材质，使传统建筑的特性得以保留。这些周到的设计确保了每个房间都能在坐拥海景和阳台的同时，其私密性也得以保护。一条带有天窗的走廊被线性布置，与各个房间相连，漫步其中，通过不同的房间，足不出户便可以领略大海的风采。

最后，走廊通向一间带火炉的起居室，那里随时欢迎着游客与奇洛埃文化的美丽邂逅。

This page, above: Entrance of the hotel. This page, below: The rooms of the hotel have patios and balconies with views to its surroundings.

本页，上：酒店入口；本页，下：酒店的房间都配有可以观景的露台和阳台。

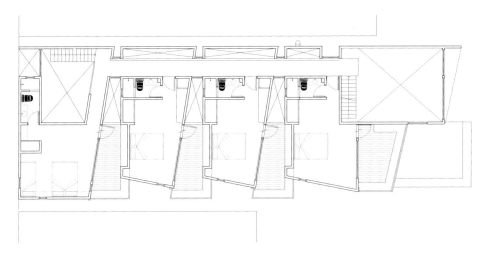

Second floor plan／二层平面图

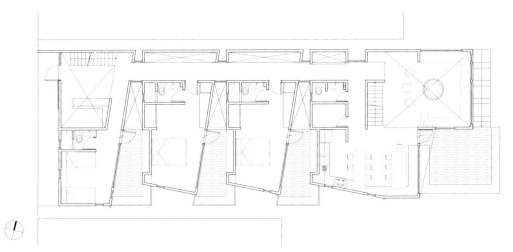

First floor plan (scale: 1/200)／一层平面图（比例：1/200）

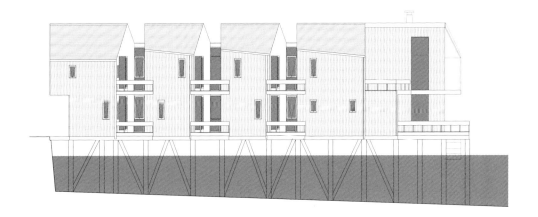

South elevation (scale: 1/200)／南立面图（比例：1/200）

Opposite: The shared living room at the end opens up to a view of the austral canals. This page: The central corridor with a linear skylight leads its guest to this communal room. Photos on p. 106 (above), p. 108 by Alvaro Vidal. Photos on , p. 106 (below), p. 109 by Pablo Casals.

对页：走廊尽头的公共起居室，南侧面向运河的景色。本页：带有天窗的直走廊尽头通向公共起居室。

Cazú Zegers Architects
Hotel of the Wind - Tierra Patagonia Hotel
Sarmiento Lake, Torres del Paine, Magallanes y de la Antártica Chilenaa 2005–2011

凯泽·塞赫尔斯建筑师事务所
风之酒店——蒂拉·巴塔哥尼亚酒店
智利南极大区麦哲伦,托雷斯德尔佩恩,萨米恩托湖 2005-2011

pp. 110–111: View of the hotel against the mountains of Torres del Paine National Park and Sarmiento Lake. pp. 112-113: A scene of the hotel at night. All photos on pp. 110–123 by Cristobal Palma unless otherwise specified.

第 110-111 页：酒店面对百内国家公园的群山和萨米恩托湖畔。第 112-113 页：酒店夜景。

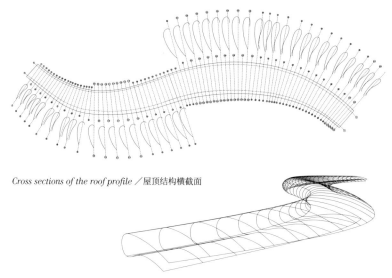

Cross sections of the roof profile ／屋顶结构横截面

3D study of roof profile ／屋顶结构 3D 研究模型

The hotel is located at the entrance of Torres del Paine National Park on the shores of Sarmiento Lake, one of the limits of this National Park in Chile. The water acts as a supporting plane for the splendid Torres del Paine massif. Its magnificence makes one think of an extended project in dialogue with the vast territory. The form seeks to merge with the metaphysical landscape, not to interrupt it. The shape of the hotel is reminiscent of an old fossil, a prehistoric animal beached on the shore of the lake, similar to those found and studied by Charles Darwin. The building is as though born out of the land, like a fold in the terrain that the wind has carved in the sand. It is anchored to the ground with stone slopes and is completely covered in washed lenga wood paneling. This finishing gives the hotel a silvery sheen, typical of old wooden houses worn away by winter. The spatial solution aims for warm and cozy spaces structured by internal pathways, allowing the building to inhabit its extension.

Credits and Data

Project title: Hotel of the Wind – Tierra Patagonia Hotel
Client: Katari S.A
Location: Sarmiento Lake, Torres del Paine, Magallanes y de la Antártica Chilena, Chile
Design: 2005–2006
Completion: 2011
Architect: Cazú Zegers
Design Team: Rodrigo Ferrer, Roberto Benavente
Project Team: Enzo Valladares and Associates (structural), George Somerhoff (acoustics), Salfa Corp / Juan Pablo Libacic (construction management), Coz Cía. Ltd. / Gustavo Soto (technical), Paulina Sir and Gaspar Arenas (illumination), Catalina Philips / Gerardo Ariztía (landscape), Carolina del Piano / María Jesús Yáñez (decoration)
Project area: 4,900 m² (built area), 70 ha (site area)

This page, above: Outdoor terraces around the hotel provide the perfect spot for a panoramic view of the mountains. Photos on p. 115 (above), p. 121 by Morten Andersen. This page, bottom: Main entrance. Photo by Pia Vergara.

本页，上：酒店周围的室外露台，是最适合观赏湖光山色的地方；本页，下：主入口。

pp. 116-117: View along the corridor leading from the reception to the living and dining hall.
第 116-117 页：连通服务台和起居室兼餐厅的走廊。

这座酒店位于萨米恩托湖湖畔，百内国家公园入口处，地处这座智利国家公园的边缘地带。萨米恩托湖如托盘一般映衬着百内国家公园的群山。壮丽的景色激发了人们建造能与大自然对话的建筑的构想。在造型上，项目寻求与当地独特地形的融合，而非唐突地介入其中。酒店静卧在湖边，它的造型仿佛在追忆着那些达尔文研究过的史前动物化石。建筑形体如同一条风在沙地里吹出的折痕，充满了自然的气息。它通过岩石的坡道与大地相连，表面全部由处理过的莲茄木板所覆盖。这层外表皮给酒店披上了一层银色的光晕，如同冬日里古旧的木屋。由各种通道构成的内部空间，温暖而舒适，并保留了酒店扩建的可能性。

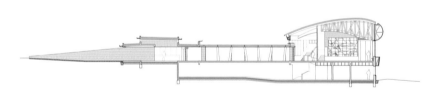

Section A ／剖面图 A

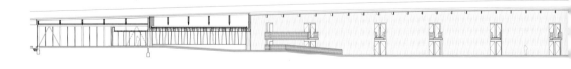

Section B (scale: 1/650) ／剖面图 B（比例：1/650）

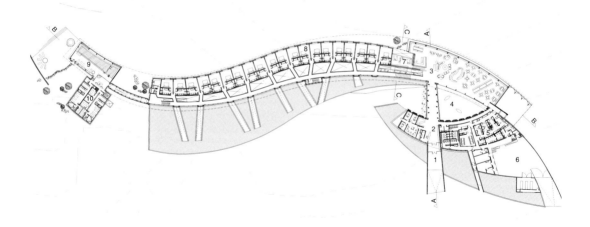

First floor plan (scale: 1/1,600) /一层平面图（比例：1/1,600）

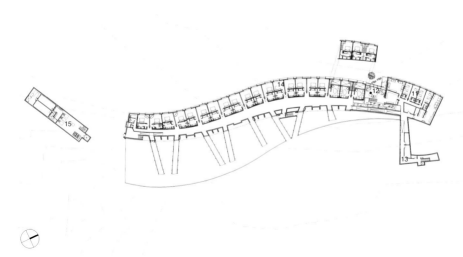

1. Main entrance
2. Reception and shop
3. Dining room
4. Patio
5. Kitchen and services
6. Unloading
7. Reading room
8. Rooms (first floor)
9. Pool
10. Spa
11. Personal room
12. Break / TV room
13. Circulation and wareho
14. Rooms (second floor)
15. Machine room

1. 主入口
2. 前台和商店
3. 餐厅
4. 露台
5. 厨房和服务
6. 卸货区
7. 阅览室
8. 房间（一层）
9. 游泳池
10. 水疗中心
11. 私人房间
12. 休息室/放映室
13. 流通和仓库
14. 房间（二层）
15. 机房

Basement floor plan /地下层平面图

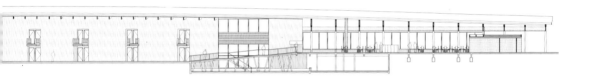

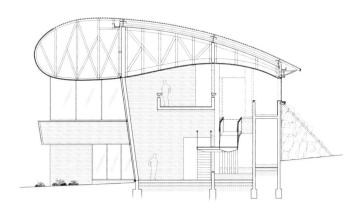

Section C (scale: 1/250) ／剖面图 C（比例：1/250）

Opposite: Corridor on the upper floor linking the private rooms to the common areas. This page: Interior view of a hotel room.
对页：连通客房和公共空间的上层走廊。本页：客房内景。

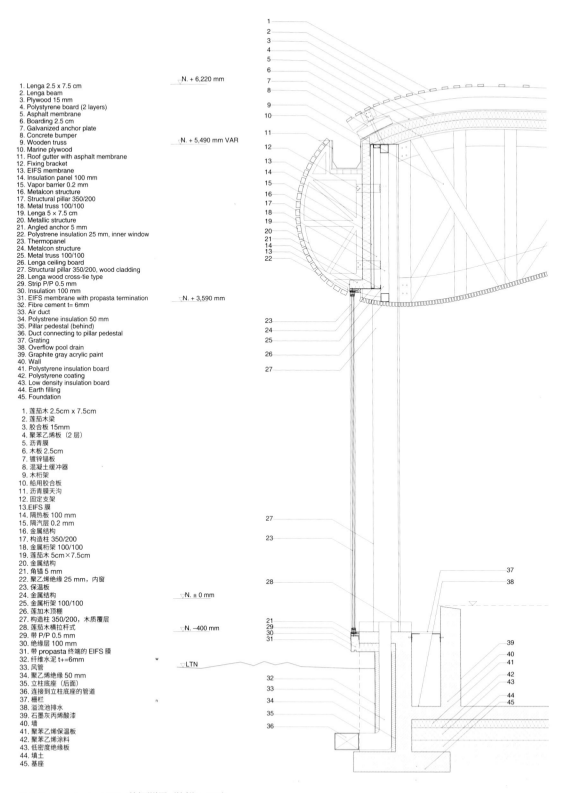

1. Lenga 2.5 x 7.5 cm
2. Lenga beam
3. Plywood 15 mm
4. Polystyrene board (2 layers)
5. Asphalt membrane
6. Boarding 2.5 cm
7. Galvanized anchor plate
8. Concrete bumper
9. Wooden truss
10. Marine plywood
11. Roof gutter with asphalt membrane
12. Fixing bracket
13. EIFS membrane
14. Insulation panel 100 mm
15. Vapor barrier 0.2 mm
16. Metalcon structure
17. Structural pillar 350/200
18. Metal truss 100/100
19. Lenga 5 × 7.5 cm
20. Metallic structure
21. Angled anchor 5 mm
22. Polystrene insulation 25 mm, inner window
23. Thermopanel
24. Metalcon structure
25. Metal truss 100/100
26. Lenga ceiling board
27. Structural pillar 350/200, wood cladding
28. Lenga wood cross-tie type
29. Strip P/P 0.5 mm
30. Insulation 100 mm
31. EIFS membrane with propasta termination
32. Fibre cement t= 6mm
33. Air duct
34. Polystyrene insulation 50 mm
35. Pillar pedestal (behind)
36. Duct connecting to pillar pedestal
37. Grating
38. Overflow pool drain
39. Graphite gray acrylic paint
40. Wall
41. Polystyrene insulation board
42. Polystyrene coating
43. Low density insulation board
44. Earth filling
45. Foundation

1. 莲茄木 2.5cm x 7.5cm
2. 莲茄木梁
3. 胶合板 15mm
4. 聚苯乙烯板（2层）
5. 沥青膜
6. 木板 2.5cm
7. 镀锌锚板
8. 混凝土缓冲器
9. 木桁架
10. 船用胶合板
11. 沥青膜天沟
12. 固定支架
13. EIFS 膜
14. 隔热板 100 mm
15. 隔汽层 0.2 mm
16. 金属结构
17. 构造柱 350/200
18. 金属桁架 100/100
19. 莲茄木 5cm×7.5cm
20. 金属结构
21. 角锚 5 mm
22. 聚乙烯绝缘 25 mm，内窗
23. 保温板
24. 金属结构
25. 金属桁架 100/100
26. 莲加木顶棚
27. 构造柱 350/200，木质覆层
28. 莲茄木横拉杆式
29. 带 P/P 0.5 mm
30. 绝缘层 100 mm
31. 带 propasta 终端的 EIFS 膜
32. 纤维水泥 t+=6mm
33. 风管
34. 聚乙烯绝缘 50 mm
35. 立柱底座（后面）
36. 连接到立柱底座的管道
37. 栅栏
38. 溢流池排水
39. 石墨灰丙烯酸漆
40. 墙
41. 聚苯乙烯保温板
42. 聚苯乙烯涂料
43. 低密度绝缘板
44. 填土
45. 基座

Detail section (scale: 1/45) ／剖面详图（比例：1/45）

This page: View of Sarmiento Lake from the outdoor pool terrace.
本页：从室外带水池的露台眺望萨米恩托湖。

Coz, Polidura, Volante, Soto
Museum of the Atacama Desert
Huanchaca Ruins National Monument, Antofagasta 1996-2009

柯兹，波里杜拉，博兰德，索托
阿塔卡马沙漠博物馆
安托法加斯塔，胡安查卡遗迹纪念碑 1996-2009

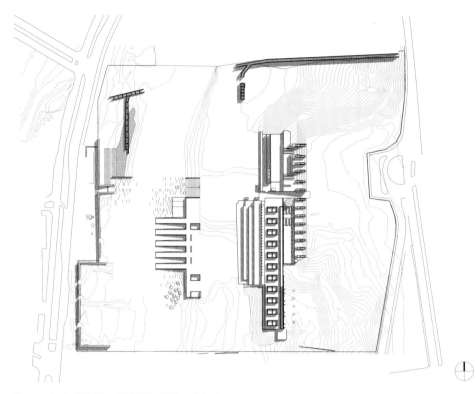

Site plan (scale: 1/4,000) ／总平面图（比例：1/4,000）

The Museum of the Atacama Desert is located at the feet of the Huanchaca Ruins, a silver foundation in the south of the city of Antofagasta, north of Chile. The ruins, which are regarded as a national monument since 1974, therefore constituting the starting point, which sets the outline for creating a strategy that was finally adopted for the museum. The main objective is to value the ruins and make them stand out by using solid masonry conforming to simple identifiable volumes and walls of monumental scale that trace the White Beach silver foundry which industrial functions ended in 1902.

Volumes and mathematical rhythms are expressed by the ruins through voids and heavy massed shapes that were translated into the geometry of the 5 rampant rooms that compose the east-west axis of the museum. At the same time these volumes connect the 2 existing terraces in the terrain. This operation is intended in order to construct a base under the monument, not interfering with the views that are directed intentionally by framing in the architecture and thus generating fluentness throughout the cultural park of Huanchaca.

The museum contains 5 permanent exhibition rooms, 2 of them are dedicated only to the geological desert of Atacama, while 2 other volumes are dedicated to the mining industry and the first northern inhabitants of Chile. The last room is dedicated to the outer space containing donations by the European Southern Observatory. On the north access, a vestibule accommodates the Huanchaca Room, a space dedicated to temporary exhibitions and activities of the museum's extension. This space visually connects and relates to the ruin through a long gash that allows views toward the pre-existing ruins and its splendor while still being inside of the museum.

3 patios resolve climatic control by creating cross ventilation, with one containing technical aspects of the museum. These patios define and visually relate the interior of the building with the ruins that stand out, reaffirming the main goal of the project. The first patio to the north, identify with the visitor through the conservation and museography work the institution has to offer. Through large window panes, the second patio expands and connects the vestibule of the 100-seater auditorium which faces the front of the cafe, souvenir shop and service area that define the museum's public area. The final patio is completely dedicated to the technical aspects of the building – concentrated and located in this space under a habitable cover with grating. This area is naturally ventilated and protected for security reasons.

The ventilations and installations of the building are grouped on a vertical concrete monolith which eliminates any element that could disturb the visual relation between the cover and the views to the ruins, endowing it with a sense of being the only element that escapes the horizontality of the building. The habitable roof delivers a platform with an open view to the monument while visually connecting the museum with the sea – reestablishing its permanent relationship between the abrupt topography presented by the terrain, and the slope located west of the Huanchaca Cultural Park.

Credits and Data
Project title: Museum of the Atacama Desert (Ruinas de Huanchaca)
Client: Fundación Ruinas De Huanchaca
Location: Huanchaca Ruins National Monument, Antofagasta Region, Chile
Design: 1996–2008
Completion: 2009
Architect: Ramón Coz Rosenfeld, Marco Polidura Álvarez, Iñaki Volante Negueruela, Eugenia Soto Cellino
Design Team: Carolina Agliati, Benjamin Ortiz, Diego Salinas, Carlos Valenzuela
Project Team: Santolaya Ingenieros Consultants (structural engineer), Mónica Perez (lighting)
Project area: 2,000 m^2 (built area), 90,000 m^2 (site area)

---

pp. 124–125: Aerial view of the museum with the Huanchaca Ruins and the city of Antofagasta in the background. pp. 126–127: The museum is geometrically split into 5 sections across the east-west axis to provide visual continuity to the volume and mathematical rhythms expressed by the ruins. Opposite: View of the museum from the Huanchaca Ruins looking out onto the Pacific Ocean. All photos on pp. 124–133 by Sergio Pirrone.

第 124-125 页：鸟瞰博物馆，背景是胡安查卡遗迹和安托法加斯塔市。第 126-127 页：博物馆沿着地形东西方向被分割成 5 个断层，使体量更好地衬托遗迹，也演绎着一种数学的韵律感。对页：从胡安查卡遗迹俯瞰博物馆，远处是太平洋。

阿塔卡马沙漠博物馆位于智利北部，安托法加斯塔市南部一座名叫胡安查卡的银矿精炼厂废墟脚下。这片废墟自 1974 年起被指定为国家历史遗产，这也是兴建这座博物馆计划的起点。这个项目的主要目标是：重现停业于 1902 年的白滩银矿精炼厂的风貌；利用坚实、简单而富有特色的体量构造，以及令人印象深刻的墙体，让发生在这里的故事广为流传。

遗迹中时而空洞，时而繁复的造型，以及组成博物馆东西轴线的 5 个巨大的房间，传达出构造的体量感和数学的韵律感。同时，这些体量与自然地形的两个高台相连，这个布局意在构建一个遗迹之下的地基，来避免干扰到建筑中预先设计好的观景方向，并串联起整个胡安查卡文化公园内的流线。

博物馆内设有 5 个常设展厅，其中两个专门用来展示阿塔卡马沙漠的地理风貌，另外两个是关于采矿文化和智利北部的首批住民。最后一个展厅展示外太空相关的内容，陈列着由欧洲南方天文台捐赠的展品。北入口的前厅设有胡安查卡厅，用于举办临时展览以及开展博物馆内的活动。视觉上，这个空间通过一个细长的切口与现存的遗迹相连，人们在博物馆内便可领略到壮丽的风光。

3 个中庭使室内空间形成了空气对流，解决了环境控制的问题，其中一个还收纳了博物馆的仪器设备。这些中庭定义了建筑的室内空间，并在视觉上将室内与室外的遗迹相连，再次强调了主题。北侧的第一个中庭，通过展示博物馆的资料保存与分类展陈，建立了与来访者之间的沟通。透过巨大的玻璃窗，第二个中庭扩大并连接着一个带有 100 个座位礼堂的门厅，门厅朝向咖啡厅、纪念品商店和服务区，博物馆的公共区域由此界定。最后一个中庭则完全用于安放设备，所有的机器设备被收纳在一个百叶罩之中，罩子上方可供人通行。这个中庭采用自然通风并因为安全问题限制对外开放。

为了排除一切屋顶与遗迹间的视觉障碍，换气和其他设备被全部安装在一个巨大的混凝土立柱上，这使它们成为这座建筑水平布局中唯一的特例。屋顶提供了一个人们可以眺望遗迹的开阔平台，并在博物馆和大海间建立起了视觉上的联系；同时，也在突然切断的地形与胡安查卡文化公园之间重构了一种永久的关联。

Opposite, above: Approach to the museum entrance. Opposite, below: View from the cafe looking towards the patio that connects across to the vestibule of the auditorium.

对页，上：从咖啡厅观看中庭，远处连接着礼堂的门厅；对页，下：博物馆入口通道。

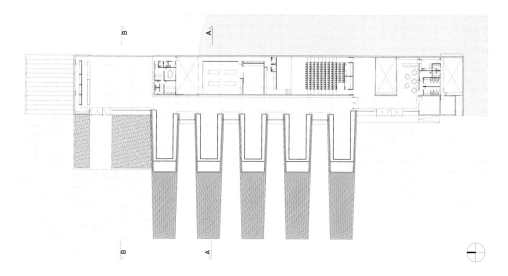

Plan (scale: 1/1,000)／平面图（比例：1/1,000）

Section A／剖面图 A

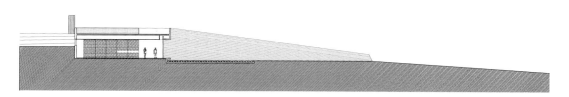

Section B (scale: 1/600)／剖面图 B（比例：1/600）

Opposite, above: A large window visually connects the interior with its external surrounding. Opposite, below: Interior view of the exhibition space.

对页，上：连结室内和室外环境的巨型玻璃窗；对页，下：展厅内景。

Undurraga Devés Architects
Retiro Chapel
Calle Larga, Valparaíso 2008-2009

翁都拉卡德维斯建筑师事务所
雷蒂罗礼拜堂
瓦尔帕莱索，卡莱拉尔加 2008-2009

Since early times religion has been a rich and constant source of inspiration for art and architecture. Interpreting the sense and form of sacred spaces is a hard task to take forward in today's secular context. This chapel, conceived in the 21st century, was understood from the initial design phases as a shelter – a space where material relinquishment, serenity, and silence inspire the pilgrims with a yearn for transcendence and consequently, an existential sense of life beyond any particular belief. Los Andes Valley, one of the most beautiful and fertile in the central area of Chile, is 70 km north of Santiago. The Carmelite Monastery of Auco is located in this agricultural land. The Retiro Chapel was built by the monastery, at the foot of the hills next to the hostel. There, a concrete volume articulates the axial character of the existing constructions with the imposing landscape of the valley. The first operation consisted of drilling the terrain, creating a cavity whose haphazard geometry was confined by a rustic stone wall. Later, 4 reinforced concrete beams are crossed to shape the cube containing the interior void. These beams extend – beyond the box – until finding support on 8 small concrete cubes located on the outer margins of the excavated space. The primitive, artisanal and tectonic world inspires its shaded interior – a box, cladded with recycled oak wood sourced from the old railway tracks, hangs from the concrete beams revealing internally a weightless box that conceals the rationality of its supports. This volume, a palpable expression of a matter free from gravity and weight, guides us to the mystery that emanates from a spiritual dimension of space. Lastly, we proposed a strategically controlled distance between the wood box and the ground, allowing the chapel to be illuminated from its lower strata, contrary to the tradition in which light, coming from above, is presented as a symbol or vehicle of the sacred. The top light is therefore restricted, keeping the interior of the box in penumbra which is further emphasized by the dark tone of the wood. It is in this chapel that the duality of the rational exterior and metaphysical interior – present in the history of religious architecture – takes on a modern expression.

Credits and Data
Project title: Retiro Chapel (Capilla del Retiro)
Client: Monasterio Carmelita
Location: Calle Larga, Valparaíso
Design: 2008
Completion: 2009
Architect: Cristián Undurraga
Design Team: Cristián Larraín (executive director), Pablo López, Jean Baptiste Bruderer
Project Team: José Jiménez / Rafael Gática (structural consultant), José Vicente Gajardo (altar consultant)
Project area: 300 m² (built area), 30,000 m² (site area)

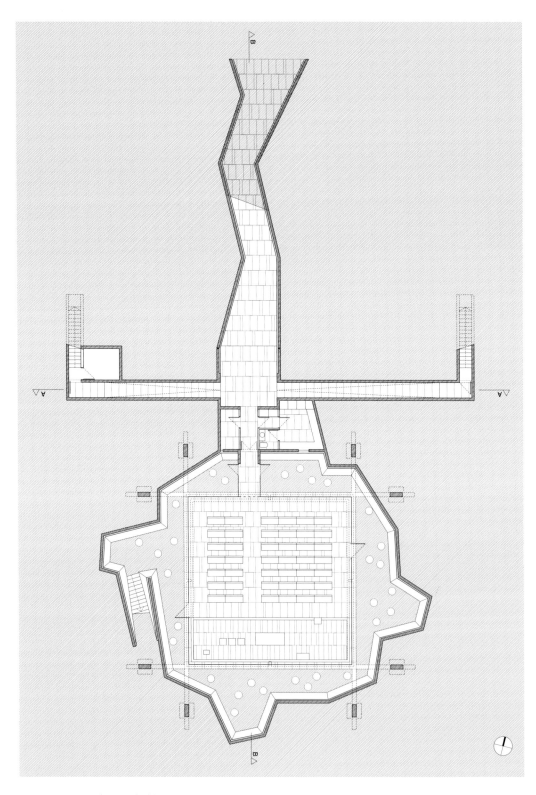

Plan (scale: 1/300)／平面图（比例：1/300）

p. 135: General view of the chapel. 4 reinforced concrete beams are crossed to shape an internal void. This page, above: Aerial view from the north. Sanctuary of Santa Teresa de Los Andes can be seen in the foreground, with the chapel in the background. This page,below: Top view. Photos on pp. 135–145 by Roberto Sáez.

第 135 页，礼拜堂全景。4 根混凝土梁相互交错围合而成的内部空间。本页，上：北鸟瞰图。近处的建筑是洛斯安第斯圣德肋撒教堂，而远处的就是礼拜堂；本页，下：顶视图。

pp. 138–139: Interior view of the chapel. The inner side of the cube is covered in recycled oak wood. Opposite: Main approach into the chapel. Photos on pp. 138-140 by Cristobal Palma. This page: View of one of the sunken walkways into the chapel. Photos on pp. 141-145 by Sergio Pirrone.

第 138-139 页：礼拜堂内景。箱型体量内侧由回收的橡木覆盖。对页：进入礼拜堂的主要通道。本页：进入礼拜堂的下沉式通道之一。

This page, right: View into the chapel from its external perimeters.
本页：从礼拜堂外部向内观看的景象

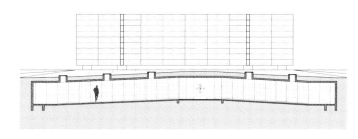

Section A／剖面图 A

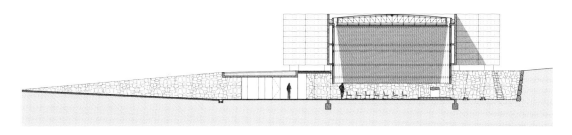

Section B (scale: 1/400)／剖面图 B（比例：1/400）

自古以来，宗教信仰就为艺术和建筑带来了丰富且永恒的灵感。在当今的世俗背景下，从感官和形式角度去诠释一个带有神圣色彩的空间变得异常困难。这座建造于21世纪的礼拜堂最初的设计概念是一个庇护所，即利用简单、安静、从容的材质引领朝圣者们超越彼此间的信仰界限，营造出一个供人感受生命的空间。洛斯安第斯峡谷位于圣地亚哥北部约70千米处，是智利中部最为美丽富饶的土地。奥科的加尔默罗会修道院就坐落于这片农业区域中。而雷蒂罗礼拜堂就坐落在修道院旁边，位于山脚的旅馆旁边。在这里，礼拜堂的混凝土体量和山谷极具特色的地形一起，将原有建筑布局的轴线表现得更加清晰。最开始的步骤，是在大地上钻挖，形成一个由天然石材围合而成的不规则空洞。接着，4根钢筋混凝土梁被置入其中，形成了一个四方形的内部空间。这些梁又继续延伸，相互交叉，最后架设在下沉空间外部边缘的8个混凝土小体量上。在原始且富有匠人手工气息的世界里，一个用于遮挡阳光的箱型空间被悬吊在混凝土梁上，给人一种空灵的氛围感。内部材料则运用收集自古旧铁路的橡木，营造出一个打破建筑构造合理性、没有重量感的空间。这个明快的空间仿佛从引力和重量中解放了一样，启发人们去探索精神世界中的神秘角落。最后，我们还设计让小结构与地面间留出一段距离，让光能够从建筑底部进入礼拜堂内。这和传统的光从上方进入不同，但也是表现神圣感的一种手法。在顶部的光受到限制的同时，又使用暗色调的木材——这都使得建筑内部处于昏暗的状态。从整体上，这座建筑以一种现代的手法，表现了宗教建筑长久以来外表理性，内在却形而上学的二重性特征。

Opposite: Close-up view of the reinforced concrete beam supported by a small concrete cube on ground level. This page: Interior view along the glass perimeter of the chapel.

对页：从近处观看钢筋混凝土梁和支撑它的小混凝土块。本页：沿着玻璃从礼拜堂外围眺望内部空间。

Emilio Marín, Juan Carlos López
Desert Interpretation Center
Ayquina, Calama, Antofagasta 2013-2015

埃米利奥·马林，胡安·卡洛斯·洛佩斯
沙漠文化中心
安托法加斯塔，卡拉马，阿伊奇纳 2013-2015

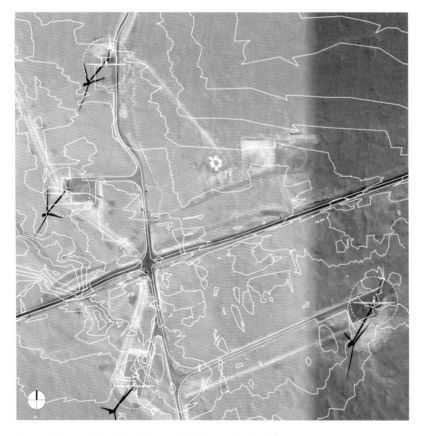

Site map showing the location of the research facility (scale: 1/6,000)
显示研究设施位置的基地图（比例：1/6,000）

The architecture brief in 2013 was set out to design a building for public use in the driest desert in the world – the Atacama Desert. The site is located close to the first wind farm in the north of Chile, between San Pedro de Atacama and Ayquina. The task was to propose a program and design a building that will be able to expose the meaning of the different layers in the desert: natural, cultural and energetic. The lack of a clear definition and focus allowed us to approach the project as an investigation into the contemporary relationship between architecture and landscape.

It is evident that the most interesting Chilean architecture of the past 20 years built its language from a dialectical relationship with the landscape. However, despite being part of the same generation of architects, we wanted to go a step further. The project aims to extend the boundaries and expand the modern vision of the interplay between architecture and landscape, where the main characters are 2 elements in a relationship of opposites. We saw a different possibility to articulate other aspects of the relationship between territory and architecture, to break the dichotomy and integrate the proposal as part of a new landscape in the desert – a device that evokes other interpretations, an observatory where visitors change their understanding of the specific natural environment.

The main strategy of the project is to integrate 3 dimensions of natural origin – geography, landscape and ecology – through 3 layers of architecture – form, material and space.

Firstly, in relation to the geographical dimension of the desert, we created different volumes, positioned against the distant volcanoes of the Andes. In the second layer, fitting the appearance of the large and disproportionate monochrome textures of the Atacama Desert, the building is covered in a single material. Corten steel envelops the whole architectural form, causing it to appear as a rock of molten steel in the vastness of the desert. The third element is the patio space which creates a new ecological dimension within the project – a new ecosystem. The volume frames a central vacuum which is protected from wind, allowing the existence of a small oasis. It is surrounded by an open corridor, which serves as a viewpoint to the sky and creates the condition for an intimate experience between the vegetation and visitors.

Credits and Data
Project title: Desert Interpretation Center (Centro de Interpretación del Desierto)
Client: Enel Green Power
Location: Ayquina, Antofagasta, Región de Antofagasta, Chile
Design: 2013
Completion: 2015
Architect: Emilio Marín, Juan Carlos López
Design Team: Alessandra dal Mos, Thomas Batzenschlager
Project Team: Cristóbal Elgueta / Macarena Calvo (garden design), Victor Palma (structural design), Hunter Douglas (facade)
Project area: 350 m²

p. 147: The building frames an internal courtyard that is protected from wind, thus allowing the existence of a small oasis. Photos on p. 147, p. 153 by Pablo Casals. pp. 148–149: The entire building is wrapped in corten steel to form an image of a rock sitting in the vastness of the Atacama Desert. Photos on pp. 148–150 by Felipe Fontecilla. Opposite: View from along the open corridor which gives its visitors an intimate experience of the internal oasis.

第 147 页：建筑物围合着中间的庭院而建，阻挡了周围的强风，使一小片绿洲得以生息。 第 148-149 页：整座建筑由耐候特种钢包裹，如同广袤的阿塔卡马沙漠中的一块岩石。对页：围着小片绿洲建造的开放回廊。

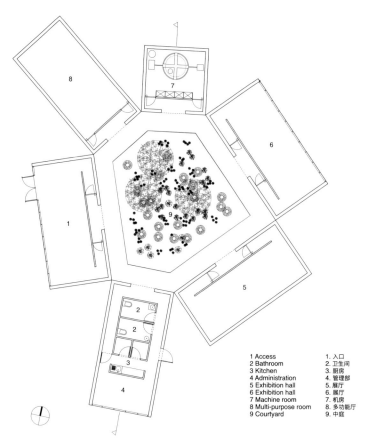

Plan (scale: 1/300)／平面图（比例：1/300）

2013年，我们开始计划在世界上最干燥的沙漠——阿塔卡马沙漠上建造一座公共建筑。项目选址于圣佩德罗德阿塔卡马和阿伊奇纳之间，在智利北部第一座风力发电厂附近。项目主题是设计一座可以展现沙漠自然、文化以及能源需求3个层次的建筑。在对项目的定义和重点尚不明确的情况下，我们开始了对现代的建筑与建筑用地间关系的探索。

实践证明20年来最有趣的智利建筑，都运用辩证法讨论了建筑表现语言和地形之间的关系。然而，虽然同属于这一代建筑师，我们却想走得更远。这个项目意在延伸并探讨：当现代建筑和建筑用地在特性上完全相反时，它们之间会建立什么样的关系。我们发现了一个产生新的其他关系的可能性，可以打破两者间的对立，成为沙漠中一道新的风景线。同时改变人们对沙漠的认知，提供一个全新的体验沙漠的场所。

设计这座建筑的主要思路是将自然的3种元素——地理、地形和生态融合进建筑的3个层面——形态、素材和空间。

首先，为了对应沙漠的地理特征，我们设计了不同的体量，分别对应远处安第斯山脉的火山群。在第二个层面，为了对应沙漠中广袤茫茫的单色调，我们将建筑用单一的素材包裹。被耐候特种钢整个包裹起来的建筑，如同一块熔化在广袤沙漠中的钢铁巨岩。第三个层面跟中庭有关，它在建筑内部形成了一个新的生态系统。多个体量围合在一起，在中央形成了一个隔绝外部强风的空间，一小片绿洲得以在此生息。绿洲被一条开敞的走廊环绕，人们可以在走廊仰望天空，或是与绿洲中的植物亲密接触。

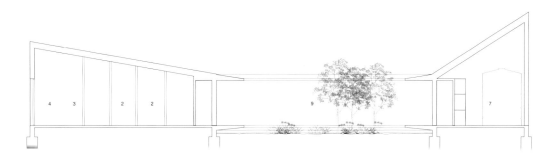

Section (scale: 1/200) ／剖面图（比例：1/200）

This page: Interior view.

本页：内景。

Max Núñez
Glass House
Pirque, Santiago 2018

马克斯-努涅斯
玻璃房
圣地亚哥-皮尔克 2018

Credits and Data
Project title: Glass House
Client: Private
Location: Pirque, Chile
Completion: December 2018
Architect: Max Núñez
Design Team: Carlos Rosas
Project Team: Osvaldo Peñaloza (structural engineer), BAS (building contractor), Juan Grimm (landscape design), Interdesign (lighting design)
Project area: 130 m²

Site plan (scale: 1/800)／总平面图（比例：1/800）

pp. 154–155: *A central pair of slanted columns, in the shape of an inverted V, holds the beam where the 2 roof vaults meet. Opposite: The glass brick roof vault reaches a maximum internal height of 6.3 m. All photos on pp. 154–165 by Roland Halbe.*

第 154-155 页：位于中央的一对倒 "V" 形斜柱，支撑着两片拱顶交汇处的横梁。对页：玻璃砖构成的拱顶让室内最大高度达到了 6.3 米。

The work of John Claudius Loudon, a Scottish botanist and garden designer, dates back 2 centuries. In *The Green-House Companion* (1824), he offers construction and planting advice to specialists and amateurs alike. Although Loudon's suggestions were not taken into consideration by Max Núñez when designing his latest building, a greenhouse in Pirque, Chile, some of his ideas reveal that the purpose of glass structures has not changed much. Loudon wrote, "An appendage to every villa, and to many town residences; not indeed one of the first necessities, but one which is felt to be appropriate and highly desirable, and which mankind recognise as a mark of elegant and refined enjoyment." Distinctly, the objective of this example in Pirque is to provide pleasure to the plant enthusiast who commissioned the project.

The Pirque greenhouse has a floor plan of 11.40 x 11.40 m, which – when seen from the outside in elevation – "floats" 70 cm above the ground. This plinth-like volume is accessible by following a stone path that brings the visitor to the main entrance. Once inside, there is a 1.30 m wide perimeter promenade which leads to a corner stairs that brings the visitor down into an inscribed squared garden of 8.20 x 8.20 m. There are 4 symmetrically arranged sliding panes of windows adjacent to each of the 4 corners of the building; and inside, 4 sets of column and slanted-column structural elements are situated on the other side of each corner. There is a central pair of slanted columns that draw an inverted V and that hold the beam where the 2 roof vaults meet. These double-vaulted roofs of glazed brick reach a maximum internal height of 6.30 m.

To keep the temperature and humidity stable throughout the year, an integrated system of heating, ventilation, and irrigation was designed. In the lower perimeter, under the elevated circulation and hidden behind a vertical grid, there is air equipment that injects heat in the winter and helps with the recirculation of air in the summer. In addition to the air equipment, the interior is ventilated by 2.45 m high windows along the flat faces of each vault; these windows open automatically when the interior exceeds 20°C. For the control of its interior humidity, a micro system of sprinklers was installed that generates a cloud of moisture. All of these technical details exist for the sole purpose of growing plants, turning this greenhouse into a captive piece of tropical garden located in the semi-arid climate of Central Chile.

Paul Scheerbart's ideas, written nearly a century after Loudon's, are an inevitable reference. In Glasarchitektur (1914), he described a development of culture that would take place in conjunction with a change of architecture, that is when people switch from an environment of closed rooms to a glazed one that allows light in completely. In his vitreous manifesto, Scheerbart wrote: "In introducing glass architecture, it is best to begin with the garden; every owner of a large garden will want to have a glass garden house." What now remains to be seen is whether or not – when the tropical forest is consolidated inside of this jewel box – the owner will clear his way in with a machete, install a hammock, and test a new domesticity.

Hearth of Glass, Daniel Talesnik,
Curator of the Architecture Museum of the TU München

Opposite: Close-up view of the technical details on its exterior.
对页：外部的技术细部近景。

约翰·克劳迪厄斯·劳登,一位活跃于两个世纪之前的苏格兰植物学家和园林设计师。在他所著的《温室导引》(1824 年)一书中,他向专业人士和业余爱好者们提供了许多关于温室构造和种植技术的建议。虽然马克斯·努涅斯在设计他的最新作品——这座位于智利皮尔克市的温室时,并未参考劳登的意见,但他的一些想法,包括玻璃的构造还是出现在了这座温室中,并未改变很多。劳登写道:"(玻璃温室)作为所有庄园以及大部分城市住宅的附属物,虽然不是必不可少的,但人们对它的需求却非常迫切,拥有温室被当成了一种优雅高贵的享受。"很明显,建造这座位于皮尔克市的模范性温室,是为了给身为植物迷的业主带来快乐。

皮尔克市温室的平面是一个 11.4 米见方的正方形。从外面来看,它仿佛"漂浮"在距离地面 70 厘米的半空中。通过一条石头铺成的小路,人们被引导到这个底座般体量的主入口。进入内部,一条宽 1.3 米的散步小道通向角落处的楼梯,人们通过楼梯可以到达一个 8.2 米见方的正方形下沉花园。4 扇推拉式玻璃窗被对称地安装在建筑 4 个角落的一边;而 4 组立柱和斜柱组成的结构则位于各个角落的另一边。在室内中央有一组 V 形的斜柱,它们支撑着两片拱顶交汇处的横梁。而这两片由玻璃砖组成的拱顶,将室内的最大高度提高到了 6.3 米。

温室内的综合系统涵盖了保温、换气和灌溉的功能,使得一年四季室内的温度和湿度保持在一定范围内。空调设备被安装在架空回廊的下方,并被垂直的格栅覆盖。它们在冬天为室内注入暖气,夏天则起着循环空气的作用。除了空调设备之外,沿着拱顶直线边缘、距离室内地面 2.45 米高的地方还安装了高窗。在室内温度超过 20°C 时,它们会自动开启并通风。为了控制湿度,建筑内安装了一个由小型喷头组成的微型系统,可以在必要时制造水雾。以上所有技术细节的目的只有一个——让植物更好地生长,从而将这座位于智利中部半干旱地区的温室,变成一座独具魅力的热带植物园。

在将近一个世纪之后,保罗·歇尔巴特在他所著的《玻璃建筑》(1914 年)一书中再次引用了劳登的观点。书中认为,随着建筑技术的发展,当人们从封闭房间的环境来到另一个能让光完全进入的房间中时,人类文明也将迎来新的进步。在保罗关于玻璃的宣言中,他写道:"关于推广玻璃建筑,最好的方式莫过于从花园入手,每个拥有大花园的人都想拥有一座玻璃温室。"试想当花园的主人面对这样一幅场景,玻璃温室仿佛透明的珠宝盒,里面盛满了热带雨林,他一定会亲自动手,在室内装上一副吊床,开始享受新的家庭生活。

《玻璃的熔炉》,丹尼尔·特雷斯尼克
慕尼黑建筑博物馆馆长

pp. 160-161: *Approach to the Glass House. This greenhouse is designed to be a captive piece of tropical garden located in the semiarid climate of Central Chile. Opposite: View of the west facade at night.*

第 160-161 页:通向温室的道路。这座温室的目的是在智利中部半干旱地区建造一座独具魅力的热带植物园。对页:西立面夜景。

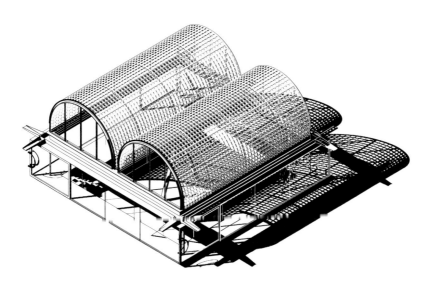

Axonometric drawing ／轴测图

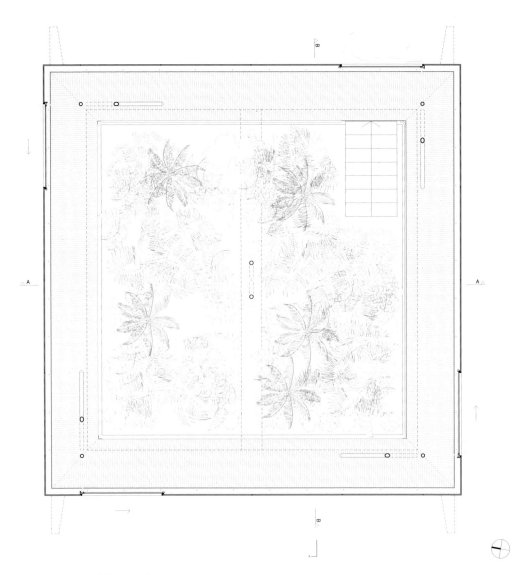

Plan (scale: 1/100) ／平面图（比例：1/100）

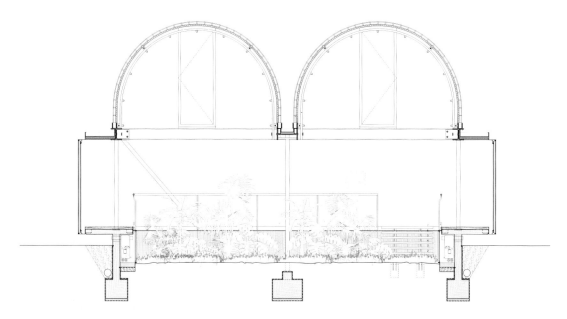

Section A ／剖面图 A

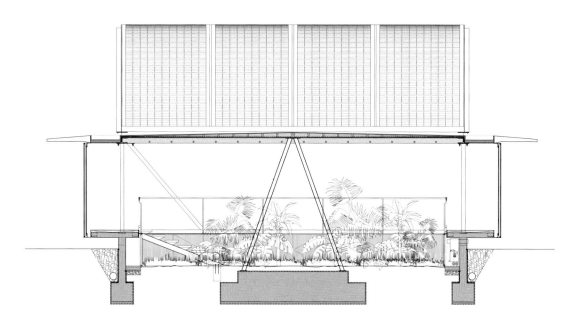

Section B (scale: 1/100) ／剖面图 B（比例：1/100）

Undurraga Devés Architects
Ruca Dwellings
Huechuraba, Santiago 2009-2011

翁都拉卡德维斯建筑师事务所
卢卡住宅
圣地亚哥，韦丘拉瓦 2009-2011

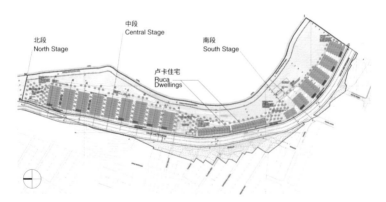

Site plan (scale: 1/2,000) ／总平面图（比例：1/2,000）

pp. 166–167: General view along one of the housing clusters. This page: Interior view of a housing unit. Photos on pp. 166–167, 170 (right) by Guy Wenborne. Photos on p. 168, 170 (left) by Pilar Undurraga.

第 166-167 页：住宅群概观。本页：居住单元内景。

The Spanish word, Mapuche means "people from the land". When the Europeans arrived in Chile, the Mapuches lived in the south central, a territory where, in harmony with nature, they developed an agriculture-based economy. Unlike other PreColumbian cultures, such as those of the north central (Andes or Mesoamerica), the Mapuches have not been builders in a traditional sense. Their sacred spaces were not temples, but the mountains, forests and rivers; their houses, called rukas, were – and in many cases still are – temporary spaces, held by light structures of branches and logs. Nestled in the landscape, they degrade over time when no longer in use and return to the ground, merging into nature's circular cycle. This explains the effort that entailed in adapting the *Mapuche* culture to the contemporary urban reality.

The social housing project was subsidized by the state, so its design had to adapt to the strict rules set by the Ministry of Housing, which focused on technical aspects and living conditions, and not on the cultural singularities of the Mapuchecommunity. In this context, the architects became mediators between both worlds. Located in the Huechuraba commune, north of Santiago, the houses are all placed one after the other on a horizontal plane, their main facade facing east in observance of the ancestral tradition of opening the main door to the rising sun. The building technique combines traditional brickwork and reinforced concrete frames, expressing the correspondence between its appearance and its structural condition. The wood diagonal is a structural element used to brace the side walls. *Rügi*, a double skin of *coligüe* (chilean bamboo), covers the partition wall and the windows of these facades, filtering sunlight in. The subtle and fragmented light evokes the penumbra of the rukas, creating a time of their own, different from time that runs through the city streets. This strategy clearly differentiates the interior and exterior, which are opposite ambits in the *Mapuche* tradition, setting a contrast with the modern one where the interior space and the landscape desire to become one. Each house is of 61 m² and has 2 floors. The ground floor contains the living area and kitchen, and the upper floor, 2 bedrooms and the bathroom. The interior was left as a bare living space so that each family could choose the finishes that met their needs and means.

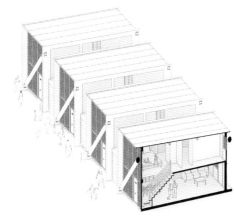

Axonometric ／轴测图

Credits and Data
Project title: Ruca Dwellings (Viviendas Mapuche)
Client: Ministry of Housing and Urban Planning, Municipality of Huechuraba
Location: Huechuraba, Santiago, Chile
Design: 2009
Completion: 2011
Architect: Cristián Undurraga
Design Team: Pablo Lopez (executive director), Raimundo Salgado
Project Team: José Jiménez / Rafael Gatica (structural consultant)
Project area: 7,130 m² (built area), 14,770 m² (site area)

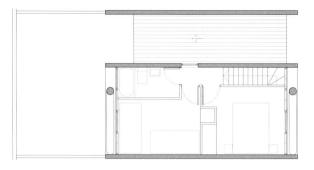

Second floor plan ／二层平面图

First floor plan (scale: 1/150) ／一层平面图（比例：1/150）

在西班牙语中，马普切人（Mapuche）的意思是"来自大陆的人们"。当欧洲移民来到智利时，这些马普切人定居在中南部地区。他们在这里与大自然和谐相处，并最终建立起一个以农业为基础的社会。和其他的前哥伦布时期中北部地区（安第斯山脉和美索美洲）的文化不同，马普切人并非传统建筑的建造者。对于他们来说，圣地不是神庙，而是于大山大河以及森林之中的存在。而他们至今还经常居住的住宅，名为"卢卡（Rukas）"，也是用简易的树枝树干搭建起来的临时建筑。这些源于大地的住宅，在废弃后便逐渐降解，最后完全回归大自然的生态循环。从这个角度也可以看出马普切文化融入当代都市的困难之处。

这个社会住宅项目由政府出资，所以它的设计必须遵循政府住宅部门的严格规定，把重点放在技术指标和居住环境上，而非马普切地区的文化特色上。在这样的情况下，建筑师扮演起了不同世界间调停者的角色。这些住宅位于圣地亚哥以北的韦丘拉瓦市。它们沿着水平的地形整齐地排成一排，主立面朝向东方，这是源自马普切人祖先们的传统——将大门朝向太阳升起的方向。建筑由传统的砖结构和钢筋混凝土组合而成，反映了建筑外观和内部构造间的关系。外部斜向安装的木柱起到了支撑墙体的作用。立面上的隔墙和窗户都被一种叫作Rügi（由智利竹子制成的双层材料）的材质所覆盖，用以阻挡部分阳光。室内微妙的光影交错唤起了人们关于卢卡的回忆，形成了一种不同于外面的城市街道，只属于住户自己的独特时空体验。这种明确区分室内外空间的手法，也与马普切文化中室内室外相互独立的传统一脉相承；而与之相对立的，则是现代建筑理论中将室内空间和周边环境融为一体的手法。每座住宅分上下两层，共61平方米。一层包含了起居室和厨房，二层则设有两间卧室和卫生间。室内几乎没有做任何装饰，以便入住的家庭根据自己的喜好和需要进行布置。

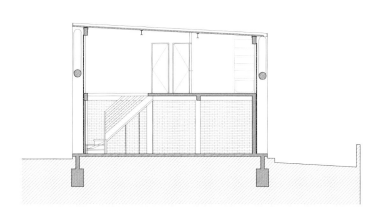

Section (scale: 1/150)／剖面图（比例：1/150）

Opposite, left: The pine wood bracing on the main façade provides the house with earthquake resistance. Opposite, right: The houses are built with a reinforced concrete frame filled in by brick walls.

对页，左：安装在主立面上、为房屋提供足够抗震性的松木柱；对页，右：住宅由钢筋混凝土框架和砖块构成的填充墙组成。

OWAR Architects
Housing Complex in Quinta Normal
Quinta Normal, Santiago 2010-2015

OWAR 建筑师事务所
金塔诺马尔集合住宅
圣地亚哥，金塔诺马尔 2010-2015

Urban renovation through strategies that incorporate higher densities, generate a positive impact on neighborhoods, and propose quality spaces for collective life, make up one of the most important challenges in cities today. These processes of urban renovation are undertaken without expelling former residents. Improving their living conditions and consolidating their social networks have become prevalent tasks when faced with high levels of segregation that reached cities like Santiago. The Housing Complex in Quinta Normal is a 4-storey affordable housing – undertaken with subsidies from the state – for 67 families in a central area of the city. The first operation is to define a central patio for all the residents, which concentrates the entrances by means of balcony-stairs that stimulate encounters. The idea is to increase public areas as a response to restrictions of the apartments and to generate a place that can accommodate collective activities. The second operation consists of liberating the first floor to situate the apartments that faces a public front away from the street-level. This elevated first floor provides a solution to parking without being a threat to common areas, and maximises the use of the entire site up to its boundary. The third operation is to generate an elevated perimeter pathway, covered by a roof, from which one accesses the vertical nuclei. This generates a limit that prevents the crossing of cars and protects the patio which is a shared, noble space of the complex.

Credits and Data
Project title: Housing Complex in Quinta Normal
Client: Housing Committee "Edificando un Sueño"
Location: Quinta Normal, Santiago de Chile
Design: 2010
Completion: 2015
Architect: Álvaro Benítez, Emilio De la Cerda, Tomás Folch
Design Team: Patricio Larraín, Agustín Infante, Juan Pablo Aguilera
Project Team: Serinco Engineering and Construction Ltd.
Project area: 4,000 m² (gross floor area), 2,500 m² (site area)

通过提高人口密度的策略实现的城市更新，对邻里社区产生积极影响，为集体生活提供优质空间。这一过程构成了当今城市最重要的挑战之一。这些城市改造过程的实施并不需要将当地原住民迁出。在圣地亚哥这样的地区，区域与区域之间的相互隔绝已经到了相当严重的程度。而那里致力于提升居住环境，加强邻里关系的项目也已经普遍存在。金塔诺马尔市的集合式住宅是一栋4层楼的公共改善住房：由政府资助建造，供中心地区的67户居民使用。设计的第一步，是在住宅的中心规划一片中庭，那里是属于所有住户的公共空间。住户们可以通过集中在中庭周围的阳台和楼梯的入口碰面，建立更好的邻里关系。这个想法的目的在于增加公寓类住宅中公共区域的面积，并创造一个可以进行公共活动的场所。第二个步骤是将人们的居住空间设在二层以上，从而把整个一层腾出来作为开放空间使用。同时，空出的一层解决了停车场对于公共区域过度占用的问题，并使得建筑用地得到了更加充分的利用。第三个步骤是设计一条覆盖有屋顶的架空环形通道，从中可以进入垂直交通核。它如同一道屏障，将车辆的活动隔绝在外，保护了集合住宅中最重要的公共区域。

Perspective section (scale: 1/250) ／剖面透视图（比例：1/250）

pp. 172–173: General view of the housing units from its courtyard. p. 174: View from along the corridor of a shared staircase. p. 175: By moving the apartments away from the public street-level, it increases privacy for its residents and provides spaces for parking without encroaching into its common areas. Opposite: Approach towards the social housing from across an intersection on the west. pp. 178: View from along the street on the south. All photos on pp. 172–179 by Erwin Brevis.

第172-173页：中庭景观。 第174页：从公共楼梯眺望中庭。 第175页：将地上一层作为停车场使用，既保护了住户的隐私，也减少了停车位对公共空间的侵占。对页：经过西面的一个十字路口处通向住宅的道路。第178页：从南侧的街道看向住宅。

Site plan (scale: 1/2,000) ／总平面图（比例：1/2,000）

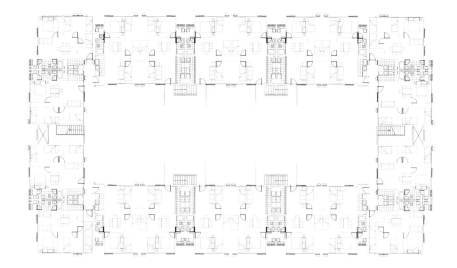

Second floor plan ／二层平面图

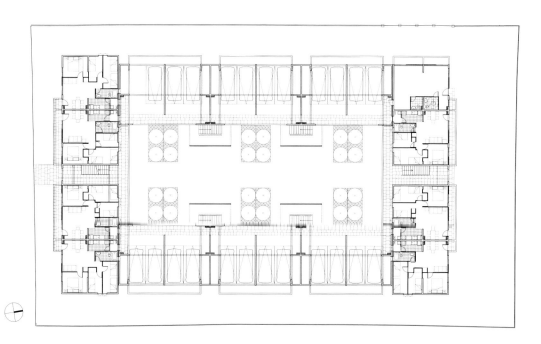

First floor plan (scale: 1/500) ／一层平面图（比例：1/500）

Izquierdo Lehmann Architects
Cruz del Sur Building
Las Condes, Santiago, 2006-2009

伊斯基尔多·莱曼建筑师事务所
克鲁兹·德尔·苏尔大楼
圣地亚哥，拉斯孔德斯 2006-2009

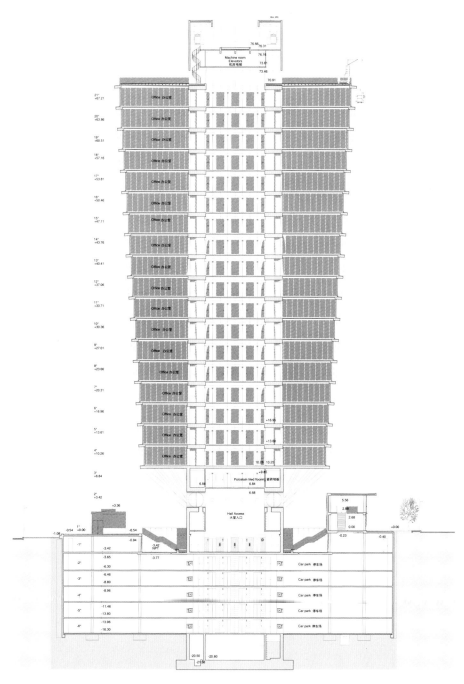

Section (scale: 1/600) ／剖面图 （比例：1/600）

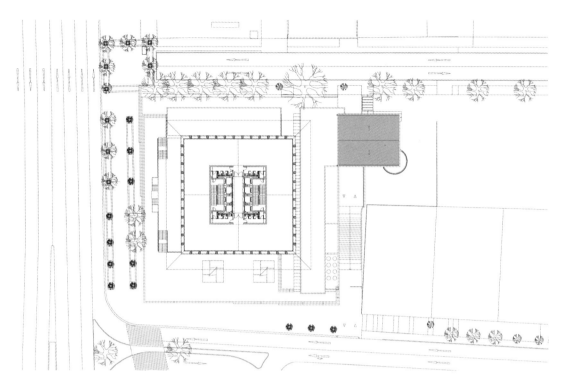

Fourth floor plan /四层平面图

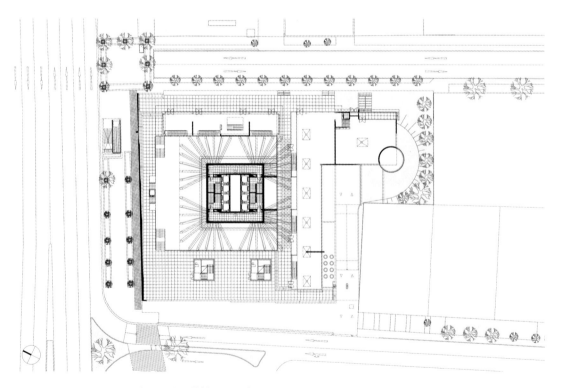

First floor plan (scale: 1/1,000) /一层平面图（比例：1/1,000）

This is an office tower with a commercial base, located at the intersection of Santiago's main axis and circular beltway. Despite its urban importance and proximity to a very busy Metro station, this place is poorly laid out – a vehicular cloverleaf surrounded by buildings of varying height and quality. The almost-square site opens on 3 fronts. Given its location, the tower faces the main avenue coming east from a mile away, serving as an iconic mark.

Given the high pedestrian density of the area, our first decision was to clear the first floor for public use. This is achieved by hiding a major part of the building's program underground, with only a corner square on the first floor enclosed. Additionally, the shaft core's capability of resisting all structural loads made it possible to structure the tower in such a way. With this solution, matching the tower to the underground columns deemed unnecessary and an optimum layout is achieved on the parking floors. Finally, the volume of the building is maximised to its permissible built area.

Floor plan surface gradually increases from floors 4th up to 21st to attain the total built area. By doing so, it reduces the shaded area produced by the bottom floor suspended over the ground; it decreases the angle of diagonal bracing supporting the series of floors; it provided column-free floor plans with increasing saleable area on upper floors; and finally, it defines the shape of a memorable landmark facing the axis of the main avenue.

Continuous windows are shaded by columns and a perimeter eave on all floors. Together with glazing setbacks, these resulted in a reduction of energy consumption up to 25% as compared to nearby buildings of the same category. Hence, the final building costs turned out to be lower than the initial estimates of similar towers. This is important because the architects believe that the economics of design purifies the rhetoric of architecture; after all, efficient use of available resources is an infallible condition for achieving beauty.

Towers are typically seen from below. Their perception varies as one approaches them. In this building, its trapezoidal shape with the distorted structural grid of its facades counteracts the fugue of lines in perspective, rendering an orthogonal volume as one gets closer. This balance plays with the relativity of the perception of volume and weight, while being conscious that the quest of stability is a decisive matter of architecture.

Credits and Data
Project title: Cruz del Sur Building
Location: Av. Apoquindo 4501, Las Condes, Santiago, Chile
Design: 2006–2007
Completion: 2009
Architect: Luis Izquierdo W.
Project Team: Santolaya Consulting Engineers
Project area: 39,408 m² (gross floor area), 3,987 m² (site area)

pp. 180–181: Aerial view of the building located in Las Condes, Santiago. p. 182: By concentrating its structural load onto the shaft core, the first floor is freed up to provide the necessary public spaces and pedestrian circulation in its dense neighborhood. Photos on pp. 180–187, 188 (above) by Cristobal Palma. Photo on p. 188 (below) courtesy of the architects.

第180-181页：鸟瞰位于圣地亚哥拉斯孔德斯的建筑。第182页：将承重全部转移到核心筒上，建筑底部可以提供更多的公共空间和步行环道，以适应周边高人口密度的环境。

This page: Interior view of a typical floor. Opposite: View from along the perimeters of the building on basement 1.

本页：标准层内景。 对页：建筑地下一层外围。

这座塔楼位于圣地亚哥市主干道和城市环道的交叉点上，上部为办公楼层，下部为商业设施。虽然这里地处闹市区，靠近一个非常繁忙的地铁站，当地的规划现状却非常堪忧：公路立交桥被四周参差不齐的楼房包围。塔楼的建筑用地几乎为正方形，与3个方向的道路相连，并面对着相隔一英里远的东面的主干道。用地的核心位置使得这座建筑成为一个重要的地标。

考虑到当地的步行人数非常之多，我们所做的第一步是将一层区域清空作为公共空间使用，将大部分的商业空间移到地下，仅在一层保留一个玻璃围合的小型广场。此外，核心筒的承重能力足以支撑整栋塔楼，这使建筑大部分悬空，仅核心与地面相连的方案得以成立。得益于这个特殊的造型，塔楼不再需要通过柱子与地基相连，同时也使得地下停车场的布置更加合理与自由。最后，这也让建筑用地的容积率达到最大。

从四层到二十七层，每层标准层的面积逐渐扩大，将建筑面积增加到了极限。低楼层面积的缩小让地面收获了更多的阳光，支撑楼层的各根结构梁之间的夹角也变得更小。其他的好处还有：楼层的空间内没有多余的柱子，更加开阔自由；高楼层的高价值办公面积得到了增加。最后，它以独特的造型面对着城市主轴线，成为一座令人难忘的地标性建筑。

每层楼都被连续的玻璃窗围合。得益于外边缘的柱子以及窗檐对阳光的遮挡，再加上窗户的缩进设计，这座塔楼比同类型的建筑物节约了25%的能耗，而建造成本也低于类似建筑的预算。这一点尤为重要，因为建筑师们相信注重经济性的理念可以为设计删繁就简，最大化利用现有的资源是实现设计美感的必要条件之一。

塔类建筑总是被人仰望，而它们给人的印象则随着人们接近方式的不同产生着多样的变化。以这座建筑来说，远看时，人们会注意到它独特的梯形外观和扭曲的网格状立面，而当人们走近时，向下收束的立面线条又使整个建筑在近处显得十分规整。这种在体量和重量之间协调平衡的手法，让人们更加意识到稳定性对于建筑的重要。

Opposite, both images: Interior views of the glass enclosed plaza on the basement 1.
对页，上下图：地下一层被玻璃围合的广场。

Cristián Fernández Architects, Lateral Architecture & Design
Gabriela Mistral Cultural Center
Santiago Commune, Santiago 2008-

克里斯蒂安·费尔南德斯建筑师事务所，横向建筑师事务所
加夫列拉·米斯特拉尔文化中心
圣地亚哥，圣地亚哥社区　2008-

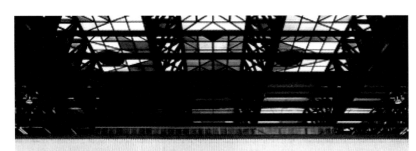

Credits and Data
Project title: Gabriela Mistral Cultural Center (UNCTAD III, 1972 / Diego Portales, 1973)
Client: Ministry of Public Works
Location: Av. Libertador Bernardo O´Higgins 277, Santiago de Chile
Design: 2008–2009
Completion: 2010 (first phase), ongoing (second phase)
Architect: Cristián Fernández Architects and Lateral Architects & Design
Design Team: (CFA) Cristián Fernández Eyzaguirre, Marcelo Fernández, Carlos Ulloa, Natalia Le-Bert, Leslie Müller, Pablo Molina / (Lateral) Christian Yutronic V., Sebastián Baraona R., Hernán Vergara H., Loreto Figueroa A., Nicolás Olate Vásquez, Nicolás Carbone, Juan Pablo Aguilera, Rodrigo Herrera, Eduardo Cid, Sebastián Bravo, Sebastián Medina, Ximena Conejeros, Irene Escobar, Ricardo Álvarez, Sebastián Bórquez, Rodrigo Carrión
Project Team: Luis Soler P & Associates (structural engineer), Musante, Astorga, Arrau Ltd. (civil engineering), Luis Farías (electrical engineer, Douglas Leonard Lighting (lighting), Enrique Montoya Engineering (sanitary engineer), Termosistemas / M & R Climatisation (air conditioning and heating), PRY Engineering (security), SIDCO (centralized control systems), SRS Ltd. (garbage disposal), Enrique Bordolini Theater Consultants (scenic design and engineering), Di. Design & Image (signage), Andrea Arenas Díaz (landscape), Mavimix Ltd. (exterior paving), Yamaimport Ltd. (audio and video), Fernández and Rojas Associates (stormwater drainage), Antonio Monasterio (intercom), Ramón López (theater consulting), Jorge Sommerhoff (acoustic), AEV Topografía (topographical survey), Musante, Astorga, Arrau Ltd. (geotechnical report), Itransporte (traffic assessment and environmental impact report), Juan Cristóbal Pérez (energy efficiency), Juan Guillermo Tejeda (artistic consulting), Patricia Squella (ADA), Engineering and Construction ICC (insulation and waterproofing), Holmes & Amaral (architectural revision), Antonio Medina (structural engineering revision)
Project area: 24,500 m² (first phase), 29,600 m² (second phase)

The Gabriela Mistral Cultural Center is one of the main cultural venues in Santiago, Chile. Located on Bernardo O'Higgins Boulevard, one of the city's most popular streets, its program was determined by the Arts and Culture Council of Chile due to an evaluation that revealed a lack of cultural infrastructure across the country. The project's main goal is to provide transparency, to open itself towards the neighboring urban fabric, and to project part of its rich and diverse interior life out onto the exterior. This way, the building becomes a relevant actor towards the promotion and broadcasting of the activities that take place within its walls. From an urban perspective, the building turns into an offering for the city – destined to become a place for arts and culture, and designed to be transparent and inclusive to invite passers-by to become part of its cultural life. Obviously, it was not the project's mission to expose everything. This was also not possible as it includes an important number of interiors dedicated to the performing arts; the challenge consists of knowing what to show, and to what degree. The building is wrapped in a system of envelopes which transparency changes gradually from being completely open to being closed and opaque. The essence of a performing arts hall resides in its ability to disconnect itself from the external reality to an absolute degree. The lights turn off... darkness takes over the room... silence sets in... and only in that moment can fantasy and a new reality unravel – the play has begun. The project's numerous performing art halls – for music, dance, theater, etc. – are designed into boxes that do not reveal their content, but offer an intuition that there is something important happening inside.

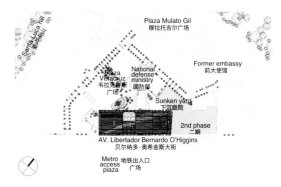

Site plan (scale: 1/5,000)／总平面图（比例：1/5,000）

p. 190, above: In 1972, this building in downtown Santiago opened as the venue for the Third United Nations World Trade and Development Conference, (UNCTAD III). Photo courtesy of Miguel Lawner, GAM Archive. p.190, below: After the coup d'etat in 1973, the cultural center turned into the seat of the new military government, who changed its name to the Diego Portales Building. Photo courtesy of El Mercurio, GAM Archive. p.191: View of the main public entrance space. Photo by Marcos Mendisabal. Opposite: General view of the cultural center from across the street. Photo on p. 192, p. 198 by Nicolas Saieh.

第190页，上：1972年，位于圣地亚哥中心区域的文化中心被作为第三届联合国贸易和发展会议的会场对公共开放；第190页，下：1973年的政变发生之后，文化中心被新的军政府接收，并被改名为Diego Portales中心。 第191页：正面主入口空间。对页：从街对面看文化中心的全貌。

加夫列拉·米斯特拉尔文化中心是圣地亚哥主要的文化设施之一。它位于城市中最热闹的街道——贝尔纳多·奥希金斯大街上。建造这座文化中心，是因为当时智利国内的文化设施数量不足，于是由政府的文化艺术委员会出面制定了这个兴建计划。项目的主要目标是将这个文化设施向周边地区开放，使其具有透明性，把丰富多彩的馆内活动投射到附近的街道中去。这样一来，这座建筑便担负起了推广和传播各种馆内活动的重要使命。从城市的角度出发，文化中心将做出巨大的贡献。它为文化和艺术而生，并被设计得富有透明感，能够吸引形形色色的路人前来，丰富大家的文化生活。当然，这个项目也并非追求彻底的对外开放，尤其在它内部已经有多个为表演艺术而设的空间时，完全的开放是不可能实现的。对设计师来说，难点在于决定要开放什么，和应该开放到何种程度。建筑被一个外表皮系统所包裹，整个系统从完全开放向闭合和不透明过渡，直至完全封闭。对于舞台艺术空间来说，重要之处在于如何彻底将内部空间从外部环境中剥离出来。当灯光暗下，黑暗和寂静接管整个空间，此时充满想象的全新世界方能得以呈现——表演开始了。为音乐、舞蹈、戏剧设置的多个大厅，被设计成一个个盒子般的空间，盒子本身的内容并不显眼，却给人一种有非常重要的事情正在内部发生的感觉。

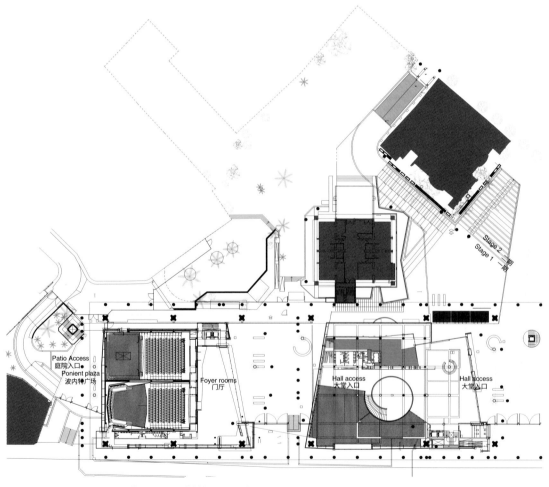

First floor plan (scale: 1/1,200)／一层平面图（比例：1/1,200）

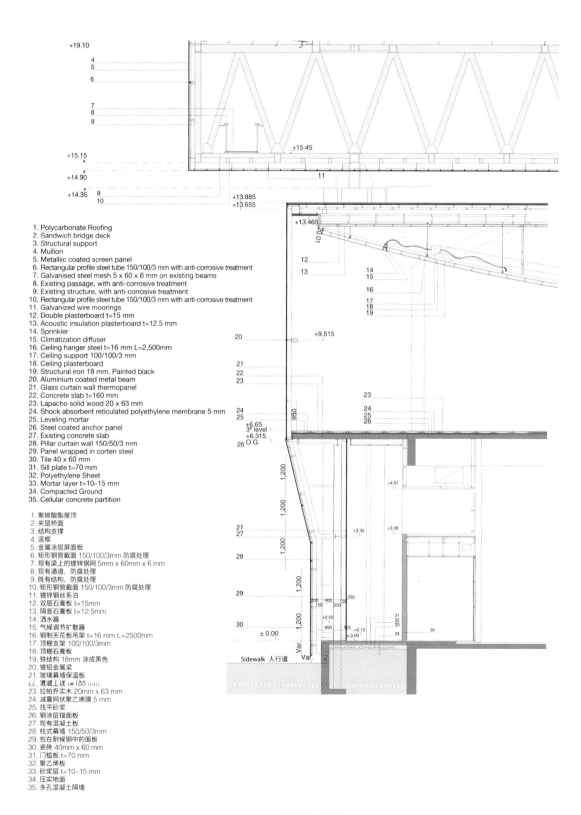

Section detail (scale: 1/120) / 剖面详图（比例：1/120）

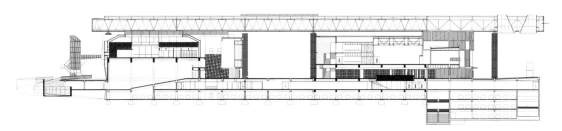

Section (scale: 1/1,200) ／剖面图（比例：1/1,200）

Opposite: View from the cafe. Corten perforated steel wrap around the walls of the building. This page: Night view from the external stairs which leads to a public outdoor stage area. Photos on p. 196, p. 197 by Juan Eduardo Sepúlveda.

对页：用在咖啡厅、建筑周围墙体的穿孔耐候钢板。　本页：夜景，室外楼梯通向一个公共的露天舞台区域。

Opposite: The corridors are gently lit by sunlight falling through the corten steel screen. This page, above: Interior view of the performing arts hall. This page, below: Stairs leading down to the smaller galleries on the basement level. Photos on p. 199 by Pedro Mutis.

对页：走廊被穿过穿孔板的阳光照亮。本页，上：表演艺术厅内景；本页，下：通向地下小型陈列室的楼梯。

Joaquín Velasco Rubio
Dinamarca 399
Cerro Panteón, Valparaíso 2013-2014

约阿奎因·贝拉斯科·鲁比奥
丹麦399号
瓦尔帕莱索,塞罗·潘特恩 2013-2014

pp. 200–201: View from along Dinamarca street. The workshop studio is seen in the foreground while the office building is in the background. This page: General view of the building on Panteón Hill. All photos on pp. 200–211 by Aryeh Kornfeld.

第 200-201 页：丹麦街的景象。近处为工作室，远处则是办公楼。本页：建在潘特恩山上的建筑外观。

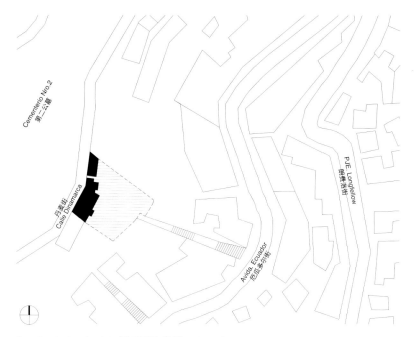

Site plan (scale: 1/3,000) ／总平面图（比例：1/3,000）

The rehabilitation of Dinamarca 399 is based on a dialogue between tradition and constructive technology, while taking into account the volumes of its historic facade of the Denmark house. This operation is expressed through a contemporary intervention using metallic steel sheets to generate flexible and luminous spaces inside the existing structure. The history of the house and its material decay are deliberately left exposed. This exposure reveals a constructive chronology. With the combination of steel and glass technology, it adapts the building's functionality to allow for new uses. Hence, the premise of our design was to proceed with our "know-how", and to intervene with readily available knowledge and technologies. As a result, the construction can still be seen as "unfinished".

Credits and Data
Project title: Dinamarca 399
Client: Inmobiliaria Meli Ltd.
Location: Dinamarca 399, Valparaíso, Cerro Panteón, Valparaíso, Chile
Design: 2013
Completion: 2014
Architect: Joaquín Velasco Rubio
Design Team: David Díaz, Angela García, Cristian Soza, Flore Brochet, Elisa Assler
Project Team: Luis Della Valle Solari (engineer)
Project area: 1,377 m² (built area), 2,045 m² (site area)

Opposite, above: View along the corridor on the first floor of the rehabilitated Denmark house. A rectangular void provides visual connectivity to the levels above and below. Opposite, below: View from the entrance hall on the first floor. This page: The different layers express the building's history.

对页，上：旧丹麦领事馆修复后的办公楼一楼走廊。矩形的挑空空间为上下层带来了视觉上的联系性；对页，下：一楼入口大厅。本页：不同层次的材质反映着建筑的历史。

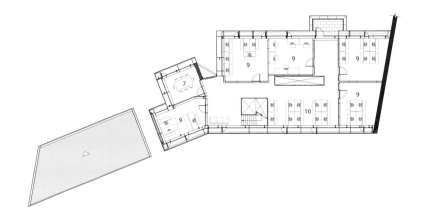

Second floor plan ／二层平面图

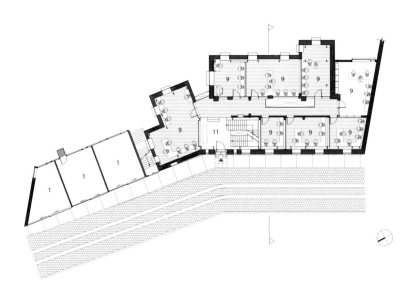

First floor plan (scale: 1/500) ／一层平面图（比例：1/500）

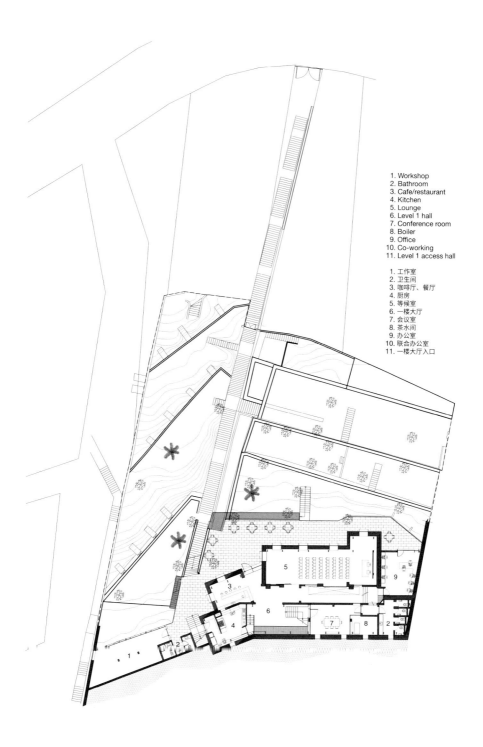

Basement floor plan ／地下层平面图

丹麦399号的改造项目是一次传统与施工技术间的对话，这次对话也考虑了旧丹麦领事馆体量中富有历史感的建筑立面。具体的做法是，将金属感浓重的现代风格板材置入现有建筑内部，产生一个灵动而明亮的空间。建筑的历史感和材料的年代感被直观地表现出来，从而记录了改建的变迁。钢铁和玻璃技术的组合使它更能胜任新的功能需要。因此，我们设计的宗旨，是利用现有的知识和技术去拓展我们对"建筑原理"的认知。从这个层面上来说，这座建筑依旧处于"未完成"的状态中。

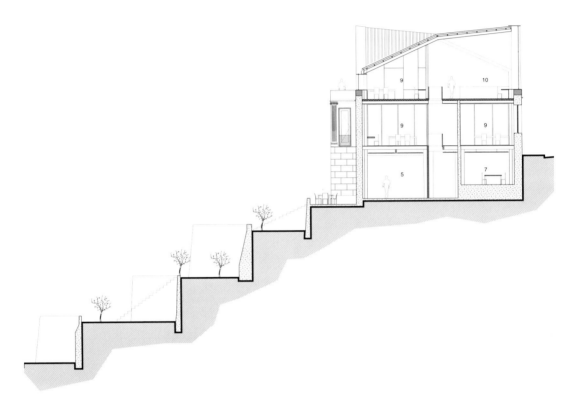

Section (scale: 1/300) ／剖面图（比例：1/300）

pp. 206–207: Interior view of the office on the second floor. Opposite: View from basement 2 of the new extension that contains the workshop studios. A staircase, seen in the background, connects to an outdoor terrace of the rehabilitated building.

第 206-207 页：二楼办公室内景。对页：位于建筑地下二层的增建部分，内部包含了工作室空间。后方的楼梯通向改造后办公楼的露台。

Cecilia Puga, Paula Velasco, Alberto Moletto
Palacio Pereira
Santiago Commune, Santiago 2012-2020

塞西莉亚·普加，宝拉·贝拉斯科，阿尔贝托·莫莱托
佩雷拉宫
圣地亚哥，圣地亚哥社区 2012-2020

In 1872, senator and businessman, Mr. Luis Pereira, commissioned French architect, Lucien Hénault, to design an urban mansion for his family. Hénault was one of many European professionals brought in by the Chilean government to design emblematic works for the new republican institutions. Educated in Ecole Nationale Superieure des Beauxarts, he developed a building of neoclassical composition and kept the facade continuous, following local tradition. However, its typology incorporated new uses and distributive systems addressing a more intricate and complex stratification of social relations. Being the element that organizes and orientates the most significant spaces, the transept constitutes as the major feature in the plan.

After being abandoned for 40 years, the Chilean state bought the palace in 2011 to transform it into the headquarters of the Ministry of Culture, Arts and Heritage. The project's material strategy sought to draw attention to the complexity of inhabiting such a structure, prioritizing neither the new intervention, nor the character of the elegant wreckage of the palace. While seeking to recover lost continuity, the project chose to celebrate the condition that existed therein at the time of starting the restoration.

The project considered a new structure to complete the original building footprint and accommodate the various offices. The vacant area resulting from previous demolitions and partial collapses was filled with a 3-dimensional, homogeneous grid of concrete pillars and beams 25 × 25 cm at a distance of 1.59 m apart. The core of this area remained free in order to rebuild the original courtyard typology creating a space capable of linking all surrounding functions. The courtyard - porous and permeable to natural light - celebrates the simultaneity, coexistence and overlapping of different periods and historical times.

Credits and Data
Project title: Palacio Pereira
Client: Directorate of Libraries, Archives and Museums
Location: Huérfanos 1501-1531, Santiago, Región Metropolitana, Chile
Design: 2012–2014
Completion: 2020
Architects: Cecilia Puga (team leader), Paula Velasco, Alberto Moletto
Design Team: Sebastián Paredes, Osvaldo Larrain, Emile Straub, Danilo Lazcano
Project Team: Alan Chandler / Fernando Pérez / Luis Cercós (restoration consultants), Pedro Bartolomé (structural engineer), Cristian Sandoval (collaborating engineer), Gabriela Villalobos / Rebecca Emmons (video and images), Alejandro Luer / Francisca Navarro (physical models), Gonzalo Puga / Claudio Cornejo (signage), Pascal Chautard (lighting), William morris (serigraphy production, pattern), Neftalí Garrido / Alejandra Jovet (serigraphy production, artists)
Project area: 2,125 m² (built area), 2,247 m² (restoration gross floor area), 4,129 m² (new extension gross floor area)

p. 213: View facing the entrance into the north wing of the original building's axis from the courtyard with the new extension. Photos on pp. 212–223 by Cristobal Palma unless otherwise specified. Opposite: Photo of the restored north wing taken during construction. Openings seen on the ground will be filled with soil and planted trees when the building reopens. This page, left: Palacio Pereira designed by architect, Lucien Hénault in 1874. Photo courtesy of the Fernández Errázuriz Family, c.1905. This page, right: Palacio Pereira, interior garden and follie. Photo courtesy of Jorge Walton from Álbum de Santiago y vistas de Chile, Santiago, 1915.

第 213 页：增建部分的中庭处通向原建筑北侧翼廊的入口。对页：改建中的北侧翼廊。当建筑重新开放时，地面上的开口处计划填土并植树。本页，左：1874 年卢西恩·亨诺设计的佩雷拉宫；本页，右：佩雷拉宫的室内花园以及装饰。

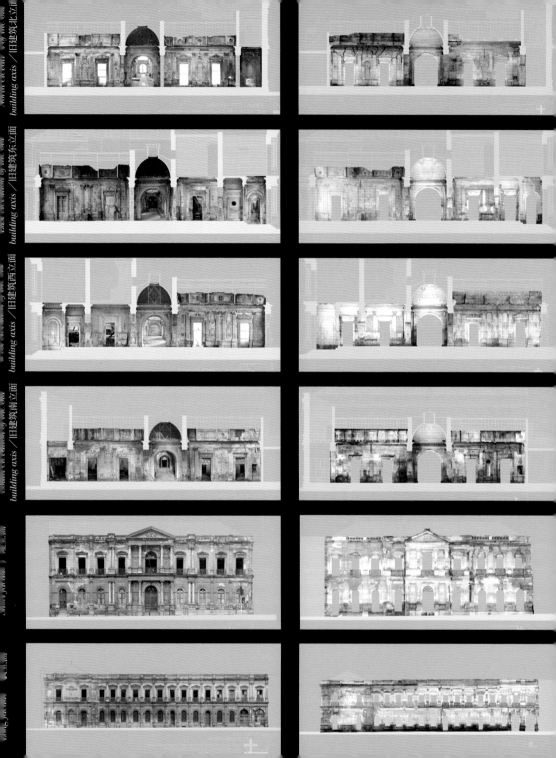

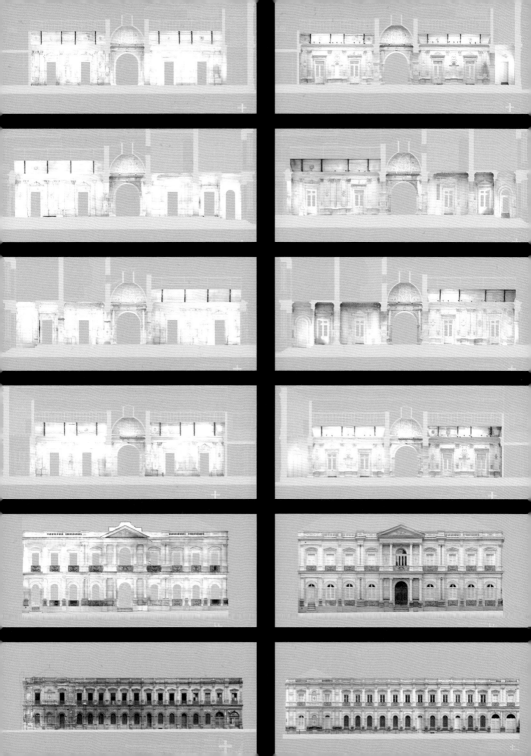

1872年，时任元老院议员，同时也是实业家的路易斯·佩雷拉先生委托法国建筑师卢西恩·亨诺为他的家族设计一座公馆。当时智利政府刚刚转入共和制，从欧洲聘请了许多专家来设计一些新的象征性建筑，卢西恩也是其中一员。曾在法国国立高等美术学院接受教育的他擅长新古典主义建筑风格。在设计这座公馆时，他遵循当地的传统，保留了建筑连续的立面。从类型学的角度上来说，他的建筑引入了新的功能和流线系统，对应了较之以往更加错综复杂的社会关系和阶级分化。从平面布局上来说，建筑内的侧廊承担着组织和配置主要空间的重要任务，也成为公馆最大的特色。

在被废弃了40年后，智利政府在2011年重新买下了这座建筑，将其作为文化、艺术和遗产部的总部。改建这座建筑的战略重点既没有放在对建筑结构加入新元素上，也没有放在刻画旧址优雅的遗迹上，而是从构造出发，力求将这座建筑内在的复杂性展现在世人眼前。将一度遗失的历史连续性慢慢找回，这个项目从改造修复开始阶段就将建筑内在的状况纳入考虑之中。

项目规划了一个新的构造用来补充原始建筑的占地面积，内部安置了各种各样的办公空间。而原来建筑毁坏崩塌所形成的空洞，则被间隔为1.59米的25厘米x25厘米（剖面尺寸）混凝土立体构架所填充。中间的区域空置，恢复了旧址原本中庭的功能，起到了连结周围各个功能区的作用。在这个充满阳光的空间里，不同的时代和历史仿佛有了同步与交错，让人浮想联翩。

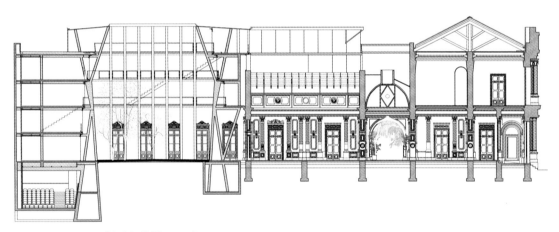

Section (scale: 1/400) ／剖面图（比例：1/400）

pp. 216-217: The restoration of Palacio Pereira was done in 3 phases. Opposite: View of the book shop. The new gold staircase acts as an element to invite people in while complementing its existing interior elements. Photo by Gonzalo Puga.

第216-217页：佩雷拉宫的修复工作分为3个阶段。对页：书店内部。黄金楼梯作为一个新的元素，邀请人们进入，同时互补其现有的内部元素。

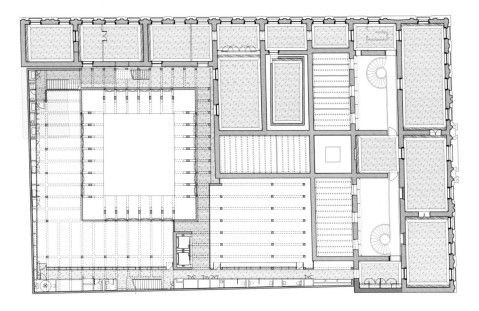

Second floor plan /二层平面图

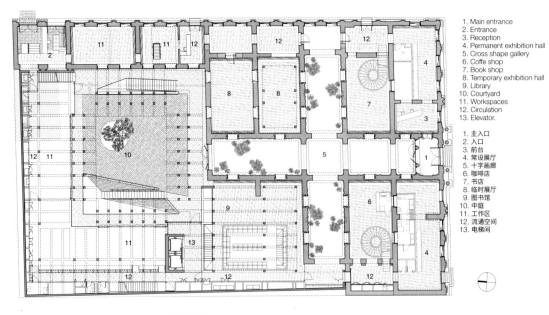

1. Main entrance
2. Entrance
3. Reception
4. Permanent exhibition hall
5. Cross shape gallery
6. Coffe shop
7. Book shop
8. Temporary exhibition hall
9. Library
10. Courtyard
11. Workspaces
12. Circulation
13. Elevator.

1. 主入口
2. 入口
3. 前台
4. 常设展厅
5. 十字画廊
6. 咖啡店
7. 书店
8. 临时展厅
9. 图书馆
10. 中庭
11. 工作区
12. 流通空间
13. 电梯间

First floor plan (scale: 1/500) /一层平面图（比例：1/500）

pp. 220-221: General view of the new extension from the courtyard. This page, top row: Restoration process of the ornaments on the walls and columns. This page, middle row: Reinforcement of the structural elements. This page, bottom row: Patterns on the wooden ceiling panels are painted by serigraphy.

第 220-221 页：从庭院看新扩建部分的全景。本页，第一排：墙壁和柱子装饰的修复过程；本页，中间一排：对结构的加固；本页，下面一排：木制天花上印刷的版画图案。

Cazú Zegers graduated as an architect from the Pontifical Catholic University of Valparaiso in 1984. She opened her own practice in Santiago, Chile in 1990 where she dedicates her time to the free exercise of her profession – developing architectural projects, industrial designs, cultural and territorial management, and research projects. Her studio's work corresponds to a "work in progress" that involves a poetic reflection on the way we inhabit territory, to find new ways from local processes and vernacular traditions. Parallel to that, she has taught at the University of Talca, Pontifical Catholic University of Chile, and University of Desarrollo.

凯泽·塞赫尔斯 1984年毕业于瓦尔帕莱索天主教大学。1990年起在圣地亚哥设立了自己的工作室，从事建筑设计、工业设计、文化和国土整治及研究项目。她的工作经常以"进行中的项目"为主题，尝试对传统做富有诗意的考察，或是从当地固有的传统中探索出新的生活方式。同时，她还任教于塔尔卡大学、智利天主教大学和德萨罗罗大学。

Cristián Fernández Eyzaguirre (1985) is an architect from the University of Chile. His professional experience includes working in the firm of Henry Smith-Miller in New York, and with Rafael Moreno in Madrid. For 10 years, he has been a principal partner of Cristián Fernández Cox. He participated in numerous presentations, conferences and seminars on the subjects of architecture, urban design and sustainability, both in Chile and abroad. Since 1998, CFA has been working on a variety of programs, from large scale private and public urban developments to small scale projects, such as an office space or furniture design. The firm has an extensive, multi-generational tradition reflected in its long professional career.

克里斯蒂安·费尔南德斯·埃扎吉雷（1985年）　建筑师，毕业于智利大学。曾就职于纽约的亨利·史密斯·米勒建筑师事务所和马德里的拉斐尔·莫雷诺建筑师事务所。也曾作为克里斯蒂安·费尔南德斯·考克斯的主要合伙人一起共事长达10年。他在智利国内外参加了许多关于建筑、城市设计和可持续性主题的演讲、会议和研讨会。1998年以来，他成立的CFA建筑师事务所涉猎广泛，小到办公空间甚至家具设计，大到大规模民间和公共城市开发项目都有参与，这同他早年丰富的职业经历是密不可分的。

Emilio Marín (1972) graduated as an architect from the Faculty of Architecture and Urbanism, University of Chile in 1998. In 2005, he founded his studio which specializes in various fields of contemporary architecture. He received the Santiago XVII Biennial Award for the work, Public Library of Licantén. He participated as a guest architect at the Porto Academy 2014 workshop in Porto, Portugal. He was part of the exhibition, LIGA 15, Mexico City. In 2019, he was selected to be the curator for Chile in the Venice Architecture Biennale 2020. Currently, he teaches as a professor at the Specialty Workshop (TES) at the Pontifical Catholic University of Chile. *Photo by Ari Espay*

埃米利奥·马林（1972年）1998年毕业于智利大学建筑与城市规划学院。2005年他成立了自己的工作室，业务涵盖了现代建筑的多项领域。他设计的利坎滕公共图书馆获得了圣地亚哥第12届双年展大奖。他作为客座建筑师参加了葡萄牙波尔图举办的2014年度波尔图研究会，也参加了举办于墨西哥城的LIGA 15展会。2019年，他被选为2020年威尼斯建筑双年展智利展的负责人。现在他作为教授任教于智利天主教大学的特别工作室（TES）。

Guillermo Acuña graduated as an architect from the Pontifical Catholic University of Chile. He owns an architectural practice based in Lo Barnechea, Santiago, Chile. His current interests are centered on researching projects which character and materiality express the possibilities of the orders in the vegetal kingdom and the traditional construction in wood and fibers – techniques that continue to manifest in the impressive geography of Chile.

吉列尔莫·阿库尼亚 建筑师,毕业于智利天主教大学。他在圣地亚哥洛巴尔内切阿拥有自己的建筑事务所。他目前的兴趣集中在研究项目上,这些项目从特点和材料性上体现了植物生态学,以及传统木材和纤维结构的可能性,这些技术在智利一些壮观的地理环境中继续发挥着作用。

Luis Izquierdo W. (photo) and **Antonia Lehmann S**. graduated from the Pontifical Catholic University of Chile (PUC) in 1980 and 1981 respectively. In 1984, they established Izquierdo Lehmann Architects in Santiago, Chile. They have built housing, educational, commercial and office buildings, single houses, interior architecture, furniture, objects, and construction systems. Many of their works have been published around the world, and presented at biennials, Harvard University GSD, and MoMA New York. They have been professors at the PUC and Andres Bello University. Their work has been distinguished with several prizes, among them, the National Prize for Architects 2004.

路易斯·伊斯基尔多·W(左图)和**安东尼娅·莱曼·S**分别在1980年和1981年毕业于智利天主教大学。1984年,他们在圣地亚哥一起成立了伊斯基尔多·莱曼事务所。业务的内容包括住宅,教育建筑,商业建筑和办公建筑,单体建筑,室内设计,家具用品以及构造系统。他们的许多作品已经被世界范围内的媒体刊登,并在各大双年展,哈佛大学GSD,纽约MoMA上展出。他们也被聘请为智利天主教大学和安德烈斯贝略大学的教授。其作品曾多次获奖,包括2004年智利建筑师奖。

Joaquín Velasco Rubio graduated from the Ponifical Catholic University of Valparaíso in 1998. In 2002, he pursued the doctorate program The Modern Form at the Polytechnic University of Catalonia (Barcelona), directed by architect Helio Piñón. Since then, he developed and collaborated on several projects in France and Spain, before settling in ValparaísoChile during 2009, where together with David Díaz and Ángela Garcia, they set up a team that specializes in low-rise multifamily housing, hotel, and commercial projects. Most of the projects are located in the historic center of Valparaíso, where the intention is to intervene in existing buildings (rehabilitation) and to design and construct new buildings.

约阿奎因·贝拉斯科·鲁比奥 1998年毕业于瓦尔帕莱索天主教大学,2002年参加了加泰罗尼亚理工大学(巴塞罗那)的建筑师赫利奥·皮涅翁领导的《现代形式》博士生项目。从那时开始,他参与了法国和西班牙的多个项目。2009年,回到瓦尔帕莱索定居,并和大卫·迪亚斯、安格拉·加西亚一起建立了一支团队,专攻低层集合住宅、酒店和商业类项目。大部分项目位于瓦尔帕莱索历史悠久的中心城区,其目的是对既有建筑的介入改造,设计并建造新的建筑。

Juan Carlos López is an architect born 1983 in Santiago, Chile. In 2013, he was recognized by Casabella magazine as one of the best worldwide architects under 30. In 2019, after 10 years of practice and teaching, he founded a studio with an artistic understanding of what an architectural object can be. Lopez worked for many years on a digital collection of shapes that inform his projects and academic work. Against the cultural simplification of our modern world, his studio pushes for architecture with a high degree of eloquence and singularity. These days, the studio is expanding into the public field, urban interventions, and collective furniture. Photo by Ari Espay.

胡安·卡洛斯·洛佩斯　1983年生于智利圣地亚哥。2013年，他被意大利《Casabella》杂志评选为世界范围内30岁以下最优秀的建筑师之一。2019年在从事了10年的实践和教育工作后，他成立了自己的工作室，研究如何对建筑进行艺术上理解。之后，洛佩斯通过数字传媒的方式向全世界分享他的项目。不同于当今世界对文化进行简化的潮流，他的工作室追求雄辩和特异的建筑风格。现在，工作室的项目已经扩展到了公共空间，城市介入和家具设计等领域。

Lateral Architecture & Design is an architectural studio founded by **Christian Yutronic** (left) and **Sebastián Baraona** (right) who both graduated from University of Chile (UChile) in 2001. Since the opening of their office, their designs and projects – such as the Gabriela Mistral Cultural Center – have won numerous awards. Together with Carlos Izquierdo, they won the first prize in the Young Architecture Competition XII of the Architecture Biennial 2002. Parallel to their practice, Yutronic has taught at UChile and University of Desarrollo, while Baraona has taught at UChile and Diego Portales University.

横向建筑师事务所是由**克里斯蒂安·尤特尼克**（左）和**塞巴斯蒂安·巴罗纳**（右）共同建立的建筑事务所。他们两人同为智利大学2001届的毕业生。自从事务所创立以来，他们设计的项目诸如加夫列拉·米斯特拉尔文化中心等已经获得了多个大奖。2002年，他们与卡洛斯·伊斯奎尔多一起赢得了智利建筑双年展"第12届青年建筑奖"的第一名。伴随着实践，他们同时从事教育工作，尤特尼克在智利大学和德萨罗罗大学任教，巴罗纳则供职于智利大学和迭戈波塔莱斯大学。

OWAR Architects is an office of architecture formed by **Alvaro Benítez** (center), **Emilio De la Cerda** (left) and **Tomás Folch** (right), who came together in 2005 and left in 2018, to focus on their individual practices. Their works explored constructive systems and typological solutions for social housing, urban design, heritage recovery, public facilities and landscape architecture.

OWAR建筑师事务所是**阿尔瓦罗·贝尼特斯**（中间），**埃米利奥·德·拉·塞尔达**（左）和**托马斯·福尔奇**（右）为专注于他们个人的实践而成立的建筑事务所。事务所成立于2005年，之后于2018年解散。他们的项目注重建造系统和分类化的解决方案，类型包括社会住宅、城市设计、古建修复、公共设施和景观建筑。

Max Núñez (1976) graduated from the Pontifical Catholic University of Chile (PUC) in 2004, and did his undergraduate studies at the Polytechnic University of Milan. In 2010, he completed his Masters in Advanced Architectural Design at Columbia University, New York. Since 2010, he has been directing Max Núñez Architects. His work was recognized with the Design Vanguard award (2017) from Architectural Recordmagazine, New York, and with the Design Award (2013, 2018, 2020) from Wallpaper magazine, London. He has also presented his works at biennials and schools in America and Europe. Parallel to his professional work, he teaches as a studio professor at PUC, where he was the director of Master in Architecture.

麦克斯·纽尔兹(1976) 2004年毕业于智利天主教大学,并在米兰理工大学完成了建筑本科课程。2010年,他在纽约哥伦比亚大学完成了高级建筑设计硕士学位,并于同年创立了麦克斯·纽尔兹建筑师事务所。所获主要奖项包括2017年纽约《建筑记录》杂志授予的"设计先锋奖",以及伦敦《卷宗》杂志2013、2018和2020年的设计奖。其作品也在美洲和欧洲的双年展和学校中展出。除了事务所的业务,他也是智利天主教大学的教授,并担任建筑学硕士课程的总负责人。

Ortúzar Gebauer Architects is an architectural practice and craft workshop in Chiloe, Chile. It is founded in 2009 by architects **Eugenio Ortúzar** (right) and **Tania Gebauer** (left) who both graduated from the University of Desarrollo and did their postgraduate in the University of Barcelona. Ortúzar has been teaching since 2002 as a workshop and urban planning teacher, and ocassionally gives talks to different universities. He gave the TED talk called, The Simplicity of the Complex in 2013. Gebauer collaborated in the Architecture Guide of Chile by DOM Publisher. Since 2019, she has been the director of Orígen Magazine which focuses on identifying the foundations of contemporary architecture in Chile.

奥图扎尔·格鲍尔建筑事务所是位于智利奇洛埃的一家专注建筑设计和施工工艺的事务所,由**尤金尼奥·奥图扎尔**(右)和**塔妮娅·格鲍尔**(左)于2009年共同设立,他们都毕业于德萨罗罗大学,并在巴塞罗那大学完成了研究生学业。奥图扎尔自2002年开始从事学校工作室以及城市规划的教育工作,并常在不同的大学发表演讲。他于2013年进行了名为《复杂中的简单》的TED演讲。格鲍尔曾为DOM出版社的《智利建筑指南》一书担当顾问。自2019年起,她开始担任聚焦当代智利建筑基础的杂志《Orígen》的主编。

Pezo von Ellrichshausen is an art and architecture studio founded in 2002 by **Mauricio Pezo** (right) and **Sofia von Ellrichshausen** (left), in Concepcion, Chile. They share the position as Associate Professor of the Practice at AAP Cornell University and have taught at Harvard University GSD, the Illinois Institute of Technology in Chicago, and the Pontifical Catholic University of Chile. Their work has been widely exhibited, namely at the Royal Academy of Arts (London), the Art Institute of Chicago, and the Venice Biennale International Architecture Exhibition. The work of the studio has been distinguished by many awards, such as the Mies Crown Hall Americas Emerge Prize by the IIT.

佩索·冯·艾奇绍森是一家从事艺术和建筑设计的事务所,由**毛乌里希奥·佩索**(右)和**索菲亚·冯·艾奇绍森**(左)于2002年在智利康塞普西翁创立。他们担任康奈尔大学的副教授职务,并同时任教于哈佛大学设计研究生院,芝加哥伊利诺伊理工学院,智利天主教大学。他们的作品在伦敦的皇家艺术院、芝加哥艺术学院和威尼斯双年展等展览上被广泛展出。他们的作品也获得诸多奖项,如伊利诺伊理工学院颁发的密斯克朗楼美洲奖新兴建筑奖。

Sebastián Irarrázaval is a RIBA International Fellow, who graduated as an architect from Pontifical Catholic University of Chile (PUC) in Santiago, and the Architectural Association in London. In 1993, he started his own practice. His works have been exhibited at various exhibitions such as the Architects College of Catalonia (COAC), the Harvard University GSD, the Milan Triennale, and many other architecture biennales. His work is published in specialized magazines, like ARQ, Casa Viva, Architectural Review, a+u, and Casabella. He has been teaching in the PUC since 1994 and is now a visiting Professor at the IUAV University of Venice, Italy.

塞巴斯蒂安·伊拉拉萨瓦是英国皇家建筑师协会的一名国际成员，毕业于智利圣地亚哥的智利天主教大学，并曾前往伦敦的建筑联盟学院进修。1993 年他创立了自己的事务所。他的作品在许多地方展出，如加泰罗尼亚建筑学院（COAC）、哈佛大学设计研究生院、米兰三年展和其他许多建筑双年展，同时也刊登于众多专业杂志，包括《ARQ》，《Casa Viva》，《Architectural Review》，《a+u》和《Casabella》。他于 1994 年起执教于智利天主教大学，现担任威尼斯建筑大学的客座教授。

Smiljan Radic´ Clarke (1965-) graduated from the Pontifical Catholic University of Chile, School of Architecture in 1989 and undertook further studies at the IUAV University of Venice, Italy. He worked with Teodoro Fernandez before opening his own firm in 1995 in Santiago, Chile. Smiljan Radic´ is the president of Fundación de Arquitectura Frágil. The aim of the foundation is to promote the study and dissemination of experimental architecture, or that of an improbable reality where the boundaries of architecture are blurred. He currently lives and works in Chile, always in close collaboration with the sculptor Marcela Correa.

史密里安·拉迪奇·克拉克 （1965-) 1989 年毕业于智利天主教大学建筑学院，之后前往威尼斯建筑大学深造。他先与特奥多罗·费尔南德斯一同工作，之后于 1995 年在圣地亚哥创立了自己的事务所。身为 Arquitectura Frágil 基金会的主席，他致力于普及、促进各种实验性建筑以及那些在建筑学范畴中看似不会发生的问题的研究。他目前定居于智利，和雕塑家玛塞拉·科雷亚进行着密切的合作。

Cristian Undurraga (1954-, photo) graduated from Pontifical Catholic University of Chile with maximum distinction and won the Young Architecture Award from the Chilean Architecture Biennale in 1977. He then founded his studio Undurraga Deves Architects, with Ana Luisa Deves in 1978. Matter, the essence of the place and planimetric clarity are the axes that mark his career. Their work is characterized by the contemporary interpretation of the historical legacy, transforming it into architecture that transcends its meaning and relavance to the present. Social, urban and cultural issues are the guiding thread of the studio's work, including a permanent structural exploration that is attentive to local possibilities and resources.

克里斯蒂安·翁都拉卡 （1954-，左图）以优异成绩毕业于智利天主教大学，并于 1997 年的智利建筑双年展上获智利青年建筑师奖。1978 年，他与安娜·路易莎·德维斯一起成立了自己的翁都拉卡德维斯建筑师事务所。物质、场所的本质以及平面上明快的轴线规划是翁都拉卡德维斯建筑风格的代表。他们作品的特点是对历史遗产的当代解读，将其转化为超越其意义并与现在相关的建筑。社会、城市和文化问题是其工作的主线，包括对当地可能性和资源的永久性结构探索。

Ramón Coz Rosenfeld (second from left) graduated from the Central University of Chile (UCEN) in 1997 as an architect. His work is exhibited at the 9th International Architecture Exhibition in Venice Biennale 2004. He teaches as a Professor at the University Mayor, Santiago, Chile. He is currently a partner and manager at Coz & Company. **Marco Polidura Álvarez** (left) graduated from the UCEN in 1998 as an architect. He was awarded by the association of Chilean architects with "Youth Promotion" in 2005. His work has been exhibited in several specialized publications in Chile and abroad. **Iñaki Volante Negueruela** (second from right) graduated from the UCEN in 1996 as an architect and completed his Ph.D. studies in ETSA Barcelona in 2000. He is the Director of the School of Architecture at University Mayor in Santiago, Chile. **Eugenia Soto Cellino** (right) graduated from the UCEN in 1995 as an architect and completed her PhD studies in ETSA Barcelona. She is a professor of a workshop in project design at the Diego Portales University and the University Mayor in Chile. Both Volante and Soto work as partner architects at REI architects.

拉蒙·柯兹·罗森菲尔德（左二）建筑师，1997年毕业于智利中央大学（UCEN），他的作品曾在2004年展于第9届威尼斯双年展国际建筑展上。他在圣地亚哥的市长大学担任教授。现在他是Coz & Company 的合伙人兼经理。建筑师**马可·波里杜拉·阿尔瓦雷斯**（左一）1998年毕业于智利中央大学，2005年获智利建筑师协会颁发的青年促进奖，其作品被刊登在国内外多家出版物上。**伊纳基·博兰德·内盖鲁埃拉**（右二）1996年毕业于智利中央大学，并于2000年在巴塞罗那建筑技术学院获得博士学位。他是圣地亚哥市市长大学建筑学院的院长。**埃乌海尼亚·索托·切里诺**（右一）1995年毕业于智利中央大学并于巴塞罗那建筑技术学院取得了博士学位。她现任迭戈·波塔莱斯大学和市长大学一项设计课题的项目教授。博兰德和索托也是REI事务所的合伙人兼设计师。

Cecilia Puga (left) is the principal of Cecilia Puga Architects based in Santiago, Chile, and currently a visiting professor in the Department of Architecture, Pontifical Catholic University of Chile (PUC). She was a member of the Executive Committee of Larrain Echenique Family Foundation (1999–2004). She is a member of the LafargeHolcim Awards jury for the Latin America region in 2020. As an independent architect, **Paula Velasco** (right) was nominated for the Rolex program Mentor and Protégé in 2014. In parallel to her professional practice, she teaches at PUC. Both Puga and Velasco often collaborate as partners in an open structure that brings together multidisciplinary teams to address different types and scales of projects. **Alberto Moletto** (center) founded Moletto Architects in 2012. He has taught at the Architectural Association, the University East London, and currently teaches at PUC. On the occasion of the Pereira Palace international competition, Moletto associates with them.

塞西莉亚·普加（左）是圣地亚哥塞西莉亚·普加事务所的负责人，现在也是智利天主教大学建筑学院的客座教授。她是拉瑞恩·埃切尼克家庭基金会（1999-2004）的执行委员之一，并担任2020年拉法基霍尔希姆奖拉丁美洲区的评委。**宝拉·贝拉斯科**（右）于2014年作为独立设计师被劳力士举办的"导师和保护人"项目委任。从事建筑设计的同时，她也在智利天主教大学任教。普加和贝拉斯科经常合作举办多学科的各种学术活动。**阿尔贝托·莫莱托**（中间）于2012年成立了莫莱托建筑师事务所，他曾在建筑联盟学院、东伦敦大学从事教育工作，现在任教于智利天主教大学。他与普加和贝拉斯科一起参加了佩雷拉宫国际竞赛。

致敬20~21世纪传奇建筑家
全解建筑世界里的光影挑战

380+ 建筑作品　　　5+家具艺术品　　　1500+摄影作品

500+ 手绘作品　100+ 论文及安藤故事　10+ 展览及建筑小品

基本信息：开本 16开/尺寸 215mm×280mm/语言 内文全中文，索引中日英三语/页码 3496页/结构 六卷/印刷 四色全彩/装
编译 安藤忠雄全集编辑部/出版发行 中国建筑工业出版社/主要作(译)者 安藤忠雄　肯尼斯·弗兰姆普顿　铃木博之　彼得·艾森

中日邦交正常化50周年纪念项目　日本国际交流基金会赞助项目

安藤忠雄全集
TADAO ANDO COMPLETE WORKS

范围 中国/内容监修 安藤忠雄建筑研究所/特别支持 日本新建筑社/书籍策划 文筑国际 IAM国际建筑联盟/主编 马卫东/执行
中川 武 三宅理一 朱涛 马卫东等/本书相关消息敬请关注官方微信公众号"安藤忠雄之家""IAM国际建筑联盟"

© 安藤忠雄建筑研究所

Spotlight:
He Art Museum
Tadao Ando Architect & Associates

特别收录：
和美术馆
安藤忠雄建筑研究所

HEM is a museum complex located in the Shunde new business district of Foshan, Guangdong Province, China. It is founded by a world-renowned entrepreneur who dedicates this project to his hometown by providing an institution that combines modern art with local and historical heritages. The museum is in the neighborhood of the business headquarter of his company, next to the city's park, which created a network for the community that puts various elements of the city at this juncture.

HEM, stands for He (Chinese character "和", to mean harmonious) Art Museum embodies its founder's wishes - to provide a harmonious life through sharing of art and culture. The character "和" has a connotation of balance and good luck, especially in the Canton area. The design of the building takes "harmony" as the theme. From the architectural design to the craftsmanship on the details, are presented through a variety of circles. This is an attempt to create a new culture center that integrates Cantonese culture by extracting the unique meaning of its geometric form in the regional context.

These circles constructed the space of the building through ripple-like expansion: from top to bottom, with the overlapping of four circles. With each area's clear-cut periphery, an enriched variation effect is created through the interaction between spaces. This is also a design adopting to the subtropical climate, in which the stark changes of light will create the emotional atmospheres.

In these circles, not only spaces are set up for exhibition and education programs, their flexibility is also ideal for the display of contemporary installations. The contrast between the circle and the square exhibition spaces has given more characteristics to HEM.

The double-helix staircase and the courtyard correspond to the overlapping circles. This dynamic presents the rich layers of the spaces that can only be achieved by the dual- Page 2 of 2 spiral design.

There are also spaces for the resting area and refreshments. Water is used as the main attractions such as the pond, which also works as a cooling device during summer. Also, the reflection of the building creates a beautiful illusion as if the basement of the building is elevated up to the sky.

I hope HEM will become a new cultural landmark in the Canton region, at the same time, a meeting point and a harmonious space for all.

Text by Tadao Ando

Credits and Data
Location: Foshan, Guangdong, China
Design: 2014.4-2017.2
Construction: 2017.2-2020.3
Structure: Reinforced Concrete
Function: Museum of art
Site area: 8,650.48 m²
Building area: 2,780.00 m²
Total floor area: 21,264.00 m²

pp. 232–233: He Art Museum of fair-faced concrete double spiral staircase. Work: Pixel Cell -- Deer #58, Kohei Nawa, composite materials, 218.6cm x 181cm x 150cm. Opposite, above: Building facade. Opposite, below: Hand drawn manuscript by Tadao Ando.

第 232-233 页：和美术馆清水混凝土双螺旋楼梯。作品：《像素细胞——鹿 #58》，名和晃平，综合材料，218.6cm x 181cm x 150cm。对页，上：建筑外立面全景；对页，下：安藤忠雄手绘稿。

这是坐落在中国广东省佛山市顺德新商务区中的一个美术馆设计项目。家族为回馈故乡文化，规划建造一座将传统历史文化与现当代艺术融汇成一体的美术馆。希望能创造出与周边设施紧密联系，跨越地域界限，充满都市艺术空间感的建筑方案。

怀着希望通过文化艺术的交流，带给人们和谐、安泰生活的夙愿，将该项目命名为"和美术馆"。建筑的设计以"和谐"为主题，从建筑设计到细部工艺，都以多样化的"圆"来呈现，尝试着创造出融汇中国岭南建筑文化的崭新艺术文化中心。

为美术馆所设计的"圆"，像水波纹一样由中心向四周扩散，构成了建筑空间的效果，同时也自然地形成了建筑形态的核心。具体而言，这些"圆"以一定的偏心率由下往上逐渐扩大，四层圆环重迭交织。立体的"圆"随之偏移，在赋予各个空间明确的中心对称的同时，更丰富了这个序列的变化效果。这样的设计也充分考虑了岭南地区亚热带气候的特性，以此营造出具有明显光影效果的建筑表情。

在圆形的建筑形态中，不但设置了中国近现代艺术展示空间、公共教育空间等非常人性化的功能区域，也设有可灵活应对当代艺术展示要求的简约立方体挑空展厅。"圆"和"方"的视觉对比，相互冲突所产生的空间差异感，为美术馆赋予了更多个性内涵。

与"圆"环迭层外观设计相呼应的，是以双螺旋楼梯为核心的五层挑空中庭设计。"圆"环构成的空间正如"圆"字所示。在富有张力的垂直空间中，以螺旋楼梯连接各层视线焦点，营造出只有"双螺旋楼梯"才能做出的层次丰富的旋转空间。

美术馆场地内的留白，可为周边商务区的人们提供休憩空间。为增加建筑亮点，景观设计以水景为主，与"圆"相呼应的水池，可作为缓和亚热带夏季酷暑的亲水装置，同时，当建筑倒映于水面，它便是建筑别具特色的底座。

希望和美术馆可以成为岭南文化的新中心，同时也是一个汇集人群，孕育"和谐"关系的场所。

安藤忠雄 / 文

Opposite: Building façade at sunset. Image by TIAN Fangfang.
对页：夕阳下的建筑外立面。

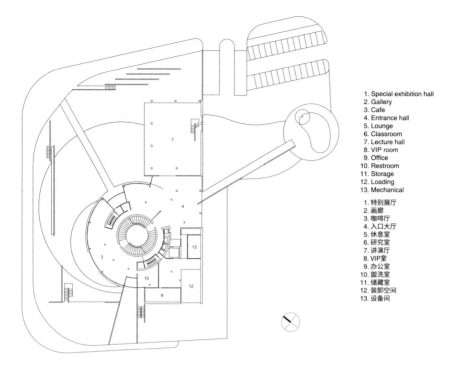

1. Special exhibition hall
2. Gallery
3. Cafe
4. Entrance hall
5. Lounge
6. Classroom
7. Lecture hall
8. VIP room
9. Office
10. Restroom
11. Storage
12. Loading
13. Mechanical

1. 特别展厅
2. 画廊
3. 咖啡厅
4. 入口大厅
5. 休息室
6. 研究室
7. 讲演厅
8. VIP室
9. 办公室
10. 盥洗室
11. 储藏室
12. 装卸空间
13. 设备间

First floor plan (scale: 1/1,500) / 一层平面图（比例：1/1,500）

Basement floor plan 2F / 地下二层平面图

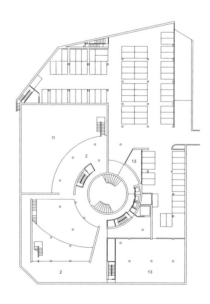

Basement floor- plan 1F / 地下一层平面图

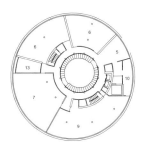

Fourth floor plan ／四层平面图

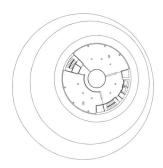

Fifth floor plan ／五层平面图

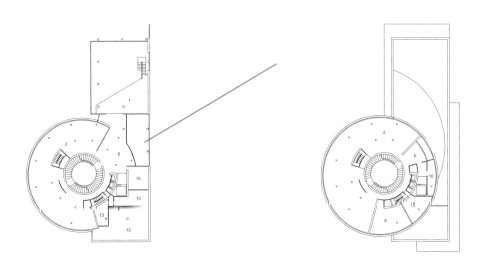

Second floor plan ／二层平面图

Third floor plan ／三层平面图

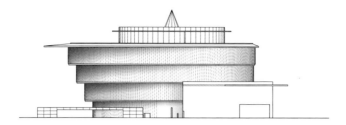

South elevation ／南立面图

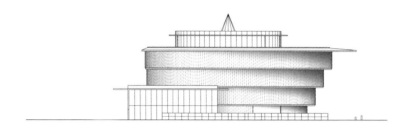

North elevation ／北立面图

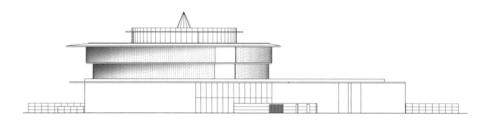

East elevation ／东立面图

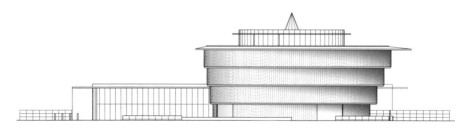

West elevation (scale: 1/1,000) ／西立面图（比例：1/1,000）

This page: Concrete facades contrast with the metal structure of the main building.

本页：混凝土外墙与建筑主体的金属结构相映成趣。

This page, Opposite: Details of building facade.

对页,本页:建筑外立面细部。

pp. 244–245: He Art Museum's fair-faced concrete double spiral staircase. This page, above: He Art Museum Hall 1. Work: Crag with Yellow Boomerang and Red Eggplant, by Alexander Calder, metal coloring. This page, below: He Art Museum Hall 2. Opposite: Stroll down the double spiral staircase to the ground floor lobby. All images on pp. 232–247 by He Art Museum except the specified.

第 244-245 页：和美术馆清水混凝土双螺旋楼梯。本页，上：和美术馆展厅一。作品：《黄色回旋镖与红茄的移动碎片》，亚历山大·考尔德，金属着色；本页，下：和美术馆展厅二。对页：从双螺旋楼梯漫步望向底层大厅。